Rethinking Urban Parks

Rethinking Urban Parks

PUBLIC SPACE & CULTURAL DIVERSITY

Setha Low
Dana Taplin
Suzanne Scheld

The University of Texas Press Austin

Photography credits: Photographs 4.2 and 4.3 are by Gabrielle Bendiner-Viani. Photographs 2.1, 3.1, 3.2, 3.3, 5.1, 5.2, 5.3, 5.4, 6.1, 6.2, 6.3, 6.4, 6.5, 8.1, and drawing 3.4 are by Dana Taplin. Photographs 1.1, 1.2, 1.3, 1.4, 4.1, and 4.4 are by Setha Low.

Copyright © 2005 by the University of Texas Press
All rights reserved
Printed in the United States of America
First edition, 2005

Requests for permission to reproduce material from this work should be sent to:
    Permissions
    University of Texas Press
    P.O. Box 7819
    Austin, TX 78713-7819
    www.utexas.edu/utpress/about/bpermission.html

♾ The paper used in this book meets the minimum requirements of ANSI/NISO Z39.48-1992 (R1997) (Permanence of Paper).

Library of Congress Cataloging-in-Publication Data

Low, Setha M.
    Rethinking urban parks : public space and cultural diversity / Setha Low, Dana Taplin, and Suzanne Scheld.— 1st ed.
        p.   cm.
    Includes bibliographical references and index.
    ISBN 978-0-292-71254-6   1. Public spaces—United States.   2. Urban parks—United States.   3. Environmental psychology—United States.   4. Multiculturalism—United States.   I. Taplin, Dana.   II. Scheld, Suzanne.   III. Title.
   HT153.L68   2005
   307.76—dc22

2005014161

Contents

	List of Illustrations	vii
	A Note on Terminology	ix
	Acknowledgments	xi
Chapter 1	The Cultural Life of Large Urban Spaces	1
Chapter 2	Urban Parks *History and Social Context*	19
Chapter 3	Prospect Park *Diversity at Risk*	37
Chapter 4	The Ellis Island Bridge Proposal *Cultural Values, Park Access, and Economics*	69
Chapter 5	Jacob Riis Park *Conflicts in the Use of a Historical Landscape*	101
Chapter 6	Orchard Beach in Pelham Bay Park *Parks and Symbolic Cultural Expression*	127
Chapter 7	Independence National Historical Park *Recapturing Erased Histories*	149
Chapter 8	Anthropological Methods for Assessing Cultural Values	175
Chapter 9	Conclusion *Lessons on Culture and Diversity*	195
	References Cited	211
	Index	219

List of Illustrations

Tables
3.1. Park-Related User Values in Prospect Park 51
3.2. Park Values in Prospect Park User Study Reclassified 52
3.3. Values and Census Group in Prospect Park User Study 53
4.1. Ellis Island Bridge Constituency Groups 72
4.2. Ellis Island: Methods, Data, Duration, Products, and What Can Be Learned 73
4.3. Value Orientations at Battery Park 82
4.4. Value Orientations at Liberty State Park 90
4.5. Neighborhood Value Orientations 96
4.6. Value Orientations: Comparison across Parks and Neighborhoods 98
5.1. Jacob Riis Park: Methods, Data, Duration, Products, and What Can Be Learned 107
7.1. Independence National Historical Park: Methods, Data, Duration, Products, and What Can Be Learned 154
7.2. Independence National Historical Park: Comparison of Cultural Groups by Content Analysis Categories 171
8.1. Qualitative Methodologies in Cultural Anthropology: Research Appropriateness 180
8.2. Constituency Analysis 181
8.3. Overview of Methods, Data, Products, and What Can Be Learned 192

Maps
3.1. Prospect Park 44
4.1 Liberty State Park and Proposed Bridge 70
5.1. Jacob Riis Park 106
6.1. Pelham Bay Park 128
7.1. Cultural Resources For African Americans 157
7.2. Cultural Resources For Asian Americans 161
7.3. Cultural Resources For Hispanic Americans 163
7.4. Cultural Resources For Italian Americans 166
7.5. Cultural Resources For Jewish Americans 168

Photos and Drawings
1.1. Shoeshine men in Parque Central in San José, Costa Rica 6
1.2. Pensioners in Parque Central in San José, Costa Rica 6
1.3. Vendors and religious practitioners in Parque Central 7
1.4. Redesigned Parque Central 7

2.1. Romantic detail—Cleftridge Span in Prospect Park   21
3.1. The Long Meadow in Prospect Park   46
3.2. Sunbathers at Prospect Park   49
3.3. Winter day, Prospect Park   49
3.4. The drummers' grove in Prospect Park   56
4.1. Circle Line ferry from Battery Park to Ellis Island   75
4.2. Battery Park landscape with Castle Clinton in the background   76
4.3. Caricatures for sale, Battery Park   76
4.4. A meadow in the northern sector of Liberty State Park   84
5.1. Jacob Riis Park bathhouse, promenade, and beach   105
5.2. Picnickers at Jacob Riis Park   113
5.3. The Clock at Jacob Riis Park   113
5.4. Park visitor cooking in shade cast by concrete wall, Riis Park   118
6.1. Promenade at Orchard Beach   129
6.2. Pelham Bay from the Orchard Beach Promenade   135
6.3. Concessions area at Orchard Beach   135
6.4. Picnicking at Orchard Beach   139
6.5. Seniors at Orchard Beach   141
8.1. Ethnographers at work at Jacob Riis Park   176

Larger maps between pp. 168-169

A Note on Terminology

During the first round of copyediting of this manuscript we tried to regularize the terminology used to refer to groups of people when described by ethnicity, race, and class. We were acutely aware that these categories are socially constructed—that is imagined, created, negotiated, and used—by people with regard to particular places, times, and circumstances, and that all labels can lead to stereotyping and essentializing of what are slippery and constantly transforming social identities. We also were concerned with how racial terms have become historically merged with notions of ethnicity and class, and how racial categories are used to justify discriminatory activities. Nonetheless, our topic was cultural diversity, and to make many of our points—which we believe to be empowering—we needed to write about people as culturally and politically relevant groups rather than as individuals, and with terminology that our interviewees and community co-workers would recognize and use to represent themselves.

Equally problematic is that each chapter is based on research conducted at different historical moments when ethnic/racial terms were shifting both within the study population (from Hispanic to Latino and from black to African American) and within the academy (from black to Afro-Caribbean American or African American). We also had problems with an unmarked "white" category, frequently used in park studies in which only the marked social category of "others" is discussed. In New York City and the Northeastern region, "white" covers many distinct ethnic and cultural groups that have very little resemblance to one another in terms of history, class status, language, and residence. For example, recently arrived Russians who use Jacob Riis Park are socially and culturally distinct from long-time Brooklyn residents in terms of their beach use and interests. As another example, we found that fourth-generation Italian Americans at Independence identified so strongly with their language and culture that they did not see the Independence Historical National Park interpretation as related to their cultural group any more than did the Puerto Rican Americans we interviewed.

In view of all these problems, we are unable to provide any fixed terminology or categories for referring to or identifying the different cultural, racial, ethnic, and class groups we discuss in this book. Instead, we relied on the categories used by the groups themselves, or employed the categories that the park managers and administrators gave us when beginning a project. Therefore, the terminology varies from chapter to chapter, and in some cases varies within a chapter if there are differences between the terms individuals use to refer to themselves and the categories that were mandated for the specific park project. Readers should not have a problem with these variations because, every day, we encounter the decision of whether to use black or African American, Latino or Puerto Rican, white or Jewish.

We hope that readers will consider the richness of this ever-changing terminology as both creative, part of the identity-making and affirming of individuals, and also destructive, in that it reflects the distinctions and dualities of black/white, white/people of color, and native/immigrant that pervade our language and can lead to discrimination in U.S. society. Although we do not focus directly on racism in the United States, racist ideology and practices underlie the cultural processes and forms of exclusion we describe in urban parks and beaches. We intend this work to be antiracist at its core, and to contribute to a better understanding of how racism, as a system of racial advantage/disadvantage, configures everyday park use and management.

Acknowledgments

The authors would like to thank the National Park Service (NPS) and especially Doris Fanelli and Martha Aikens at Independence National Historical Park, Richard Wells at Ellis Island, William Garrett at Jacob Riis Park, the late Muriel Crespi, Ph.D., past director of the NPS Applied Anthropology Program located in Washington, D.C., and Rebecca Joseph, Ph.D., and Chuck Smyth e, Ph.D., the East Coast regional directors of the ethnography program, for their support of this project. We would also like to thank the New York City Department of Parks and Recreation and the managers of Pelham Bay, Van Cortlandt and Prospect Parks—Linda Dockery, Mary Ann Anderson, and Tupper Thomas—for funding of the research reported for New York City.

Setha Low would also like to thank the staff at the Getty Conservation Institute (GCI) at the Getty Center in Los Angeles—Sheri Saperstein, Valerie Greathouse, David Myers, Kris Kelly, and Eric Bruehl—for making the writing of this book possible. A guest scholar fellowship at the GCI from January through March of 2003 enabled her to complete the first draft of this manuscript. We would also like to thank the Graduate Center of the City University of New York and particularly the Center for Human Environments and its director, Susan Saegert, for their support and assistance. Without Susan's encouragement and her staff's help, these research projects would have been much more difficult.

Some of the material in this book draws upon material published in the following articles:

Low, Setha. 2004. Social Sustainability: People, History, Values. In *Managing Change: Sustainable Approaches to the Conservation of the Built Environment,* ed. J. Teutonico. Los Angeles: The Getty Conservation Institute.

Low, Setha. 2002. Anthropological-Ethnographic Methods for the Assessment of Cultural Values in Heritage Conservation. In *Assessing the Values of Cultural Heritage,* ed. Marta de la Torre, 31–50. Los Angeles: the Getty Conservation Institute.

Low, Setha M., Dana Taplin, Suzanne Scheld, and Tracy Fisher. 2001. Recapturing Erased Histories: Ethnicity, Design, and Cultural Representation: A

Case Study of Independence National Historical Park. *Journal of Architectural and Planning Research* 18 (2): 131–148.

Taplin, Dana H., Suzanne Scheld, and Setha Low. 2002. Rapid Ethnographic Assessment in Urban Parks: A Case Study of Independence National Historical Park. *Human Organization* 61 (1): 80–93.

Taplin, Dana H. 2003. Sustainability in Urban Parks—Narrow and Broad. *Proceedings: Urban Ecology: Cities in Transition.* New York: Pace University Institute for Environmental and Regional Studies, 65–76.

Writing a book always requires aid from colleagues and friends as well. A long list of graduate students at the CUNY Graduate Center collected the data for these projects, including Charles Price-Reavis, Bea Vidacs, Marilyn Diggs-Thompson, Ana Aparicio, Raymond Codrington, Carlotta Pasquali, Carmen Vidal, and Nancy Schwartz. Kate Brower, the director of the Van Cortlandt Park project, decided not to participate in the writing of this book, but we are indebted to her for her insights and guidance. Larissa Honey and Tracy Fisher also worked on these research projects before moving on, but their work was important to our completing the projects. Comments from Matthew Cooper, the late Robert Hanna, and the seminar members at the Getty Conservation Institute—especially Randy Mason and Marta de la Torre—were particularly helpful. We would also like to thank Anastasia Loukaitou-Sideris, Benita Howell, William Kornblum, Galen Cranz, and Randy Hester for their many publications and research in this important area, and for their helpful comments.

We want to acknowledge Muriel Crespi, Ph.D., director of the NPS Applied Anthropology Program, for supporting this important work and Robert Hanna, a landscape architect who loved these parks. Both Miki and Bob died during the writing of this book, so they were never able to see the final results of their encouragement. We hope that this book will keep alive their vision of culturally vibrant and protected parks.

We are grateful that we had such excellent assistance from UT Press, especially from Editor-in-Chief Theresa May, manuscript editor Lynne Chapman, and designer Lisa Tremaine. On the CUNY side, we are grateful to Jared Becker of C.H.E.

And finally, we would like to dedicate this book to our respective partners—Joel Lefkowitz, Michele Greenberg, and Isma Diaw—in gratitude for their love and support throughout the research and writing process. It has been a long journey, and they have been incredibly helpful—from lending cars and taking photographs to cooking dinners—so that this book could be finished. Thank you to all who contributed to our work.

Rethinking Urban Parks

Chapter 1
The Cultural Life of Large Urban Spaces

Introduction

William H. Whyte set out to discover why some New York City public spaces were successes, filled with people and activities, while others were empty, cold, and unused. After seven years of filming small parks and plazas in the city, he found that only a few plazas in New York City were attracting daily users and saw this decline as a threat to urban civility. He began to advocate for viable places where people could meet, relax, and mix in the city. His analysis of those spaces that provided a welcoming and lively environment became the basis of his now-famous "rules for small urban spaces." And these rules were used by the New York City Planning Department to transform the public spaces in the city.

In this new century, we are facing a different kind of threat to public space—not one of disuse, but of patterns of design and management that exclude some people and reduce social and cultural diversity. In some cases this exclusion is the result of a deliberate program to reduce the number of undesirables, and in others, it is a by-product of privatization, commercialization, historic preservation, and specific strategies of design and planning. Nonetheless, these practices can reduce the vitality and vibrancy of the space or reorganize it in such a way that only one kind of person—often a tourist or middle-class visitor—feels welcomed. One of the consequences is that the number of open, urban public spaces is decreasing as more and more places are privatized, gated or fenced, closed for renovation, and/or redesigned to restrict activities. These changes can be observed in Latin America as well as the United States, and they are drastically reducing the number of places that people can meet and participate in public life (Low 2000).

These changes are potentially harmful to other democratic practices that depend on public space and an active public realm for cross-class and multicultural contact. At least in New York after 9/11, very few places retain the cultural and social diversity once experienced in all public spaces—but Washington Square and Union Square still do. Further, an increased defensiveness and desire for security has arisen since the terrorist attack. Concrete barriers, private guards, and police protect what were previously open spaces and buildings. The threat to public safety comes not only from the outside, but also from the danger that Americans will overreact to the destruction of the Twin Towers by barricading themselves, and denying opportunities for expressing a sense of community, openness, and optimism.

Security and Fear of the "Other"

Long before the destruction of the World Trade Center, a concern with security had been a centerpiece of the postindustrial American city, expressed in its fenced-off, policed, and privatized spaces. Although many Americans have based their concerns on a fear of the crime and violence they believe pervades cities, this antiurban sentiment is often translated into a fear of the "other" across social classes and has become a mainstay of residential and workplace segregation ever since the development of suburbs. People began moving to the suburbs to escape the insecurity of dirt, disease, and immigrant populations in the inner city as soon as trolleys made commuting feasible. And suburbs offered more than just a physical distance from the city—a more powerful social distance emerged, maintained through a complex discourse of racial stereotypes and class bias.

But even within cities, similar forms of social distance took shape. Today, for instance, wealthy New Yorkers satisfy their desire for security by living in separate zones and limited-access, cooperative apartment buildings. Other city residents rely on neighborhood-watch programs and tolerate increasing restrictions on residential behavior. Even in the face of declining crime rates, this urban fear has ended up justifying more rigid controls of urban space.

The enhanced fear of terrorism—evidenced by increasingly novel surveillance techniques—is only making it worse. New electronic monitoring tactics are being implemented across the United States. Before September 11, 2001, the prospect that Americans would agree to live their lives under the gaze of surveillance cameras or real-time police monitoring seemed unlikely. But now some citizens are asking for outdoor cameras to be installed in places like Virginia Beach to scan faces of people at random, cross-checking them with faces of criminals stored in a computer database. Palm Springs is wiring palm trees with electronic eyes on the main business street. What were once considered Big Brother technologies and infringements of civil liberties are now widely treated as necessary for public safety—with little, if any, examination of the consequences. What is at stake is the cost we are paying for this increased security, measured not just in salaries of increasing numbers of police officers or in retinal-scanning technologies, but also in the loss of freedom of movement

THE CULTURAL LIFE OF LARGE URBAN SPACES

and the cultural diversity in public space that has been so characteristic of the American way of life.

Globalization and Increased Diversity

With increasing globalization this trend has intensified. Two countervailing processes are occurring. Large numbers of people are moving from developing countries to more developed regions to obtain better jobs and education and increasingly use the public spaces of the city. Yet while the macroenvironment is becoming more diverse because of increased flows of immigrants, differences in local population growth rates, and an overall "browning" of America, local environments are experiencing increased vernacularization and homogeneity—immigrant enclaves are growing in the city, and gated communities are developing in the suburbs and edge cities. In this historical era of cultural and ethnic polarization, it has become increasingly important to engage in dialogue about these changes. How can we continue to integrate our diverse communities and promote social tolerance in this new political climate? One way, we argue, is to make sure that our urban parks, beaches, and heritages sites—those large urban spaces where we all come together—remain public, in the sense of providing a place for everyone to relax, learn, and recreate; and open so that we have places where interpersonal and intergroup cooperation and conflict can be worked out in a safe and public forum.

In 1990 Setha Low, with the help of Dana Taplin and Suzanne Scheld, founded the Public Space Research Group (PSRG) within the Center for Human Environments at the Graduate School and University Center of the City University of New York to address these issues. PSRG brings together researchers, community members, and public officials in a forum of integrated research, theory, and policy. The group provides a theoretical framework for research that relates public space to the individual, the community, and to political and economic forces. PSRG is concerned with the social processes that make spaces into places, with conflicts over access and control of space, and with the values and meanings people attach to place.

In our 15 years of studying cultural uses of large urban parks and heritage sites, we have observed the local impacts of globalization: more immigrants, more diversity, new uses of park space, less public money for operations and maintenance, and greater sharing of management responsibility with private entities. We have also witnessed responses and reactions to these changes such as efforts to reassert old-order values through historic preservation and to impose greater control over public spaces through surveillance and physical reconstruction. We have documented how local and cultural misunderstandings can escalate into social problems that threaten the surrounding neighborhoods, triggering the same processes that we have seen occurring in small urban spaces. Immigrants, in some ways the mainstay of the U.S. economy, after 9/11 have become the "other" who is feared. Restrictive management of large parks has created an increasingly inhospitable environment for immigrants, local ethnic groups, and culturally diverse behaviors. If this trend continues, it will eradicate the last remaining spaces for democratic practices, places where a wide variety of people of different gender, class, culture, nationality, and ethnicity intermingle peacefully.

Lessons for Promoting and Managing Social and Cultural Diversity

Based on our concern that urban parks, beaches, and heritage sites might be subjected to these same homogenizing forces, we began a series of research projects to ascertain what activities and management techniques would encourage, support, and maintain cultural diversity. These projects produced a series of "lessons" that are similar to William H. Whyte's rules for promoting the sociability of small urban spaces, but in this case, these lessons promote and/or maintain cultural diversity. Each lesson was derived from one or more of our park ethnographies and will be illustrated in the following chapters.

These lessons are not applicable in all situations, but are meant to provide a framework and guidelines for culturally sensitive decision making in park planning, management, and design. They can be summarized in the following six statements:

1 If people are not represented in historical national parks and monuments or, more importantly, if their histories are erased, they will not use the park.
2 Access is as much about economics and cultural patterns of park use as circulation and transportation; thus, income and visitation patterns must be taken into consideration when providing access for all social groups.
3 The social interaction of diverse groups can be maintained and enhanced by providing safe, spatially adequate territories for everyone within the larger space of the overall site.
4 Accommodating the differences in the ways social class and ethnic groups use and value public sites is essential to making decisions that sustain cultural and social diversity.
5 Contemporary historic preservation should not concentrate on restoring the scenic features without also restoring the facilities and diversions that attract people to a park.

6 Symbolic ways of communicating cultural meaning are an important dimension of place attachment that can be fostered to promote cultural diversity.

These lessons for promoting and sustaining cultural diversity in urban parks and heritage sites are just a beginning. More research and experimentation will be needed to fully understand the importance and difficulties of maintaining vibrant public spaces. But at the very least, the lessons demonstrate how diversity can be an essential component of evaluating the success of any human ecosystem. The remainder of this chapter discusses the theoretical and the practical rationales for our position. We feel it is not enough to assert that cultural and social diversity is critical to large urban sites; the argument needs to be substantiated by current social theory and practice. There are economic as well as ethical reasons for considering diversity as essential to the success of any urban place. This chapter lays the groundwork for explaining why it is so critical to planning, designing, and managing large urban spaces in the future.

Theoretical Framework
Social Sustainability

What do we mean by "social sustainability"? Following David Throsby's (1995) discussion, sustainability refers to the evolutionary or lasting qualities of the phenomena, avoidance of short-term or temporary solutions, and a concern with the self-generating or self-perpetuating characteristics of a system (Throsby 1995). Drawing a parallel with natural ecosystems that support and maintain a "natural balance," "cultural ecosystems" support and maintain cultural life and human civilization (Throsby 1999a, 1999b). Sustainable development is the preservation and enhancement of the environment through the maintenance of natural ecosystems, while culturally sustainable development refers to the preservation of arts and society's attitudes, practices, and beliefs.

Social sustainability is a subset of cultural sustainability; it includes the maintenance and preservation of social relations and meanings that reinforce cultural systems. Social sustainability specifically refers to maintaining and enhancing the diverse histories, values, and relationships of contemporary populations. But to truly understand social sustainability, we need to expand Throsby's analysis by adding three critical dimensions:

1. PLACE PRESERVATION

Cultural ecosystems are located in time and space—for a cultural ecosystem to be maintained or conserved, its place(s) must be preserved (Proshansky, Fabian, Kaminoff 1983; Low 1987). Cultural conservation and sustainability

Figure 1.1. Shoeshine men in Parque Central in San José, Costa Rica

Figure 1.2. Pensioners in Parque Central in San José, Costa Rica

require place preservation. This rather obvious point is crucial when dealing with the material environment and issues of cultural representation.

2. CULTURAL ECOLOGY THEORIES

Anthropologists employ a variety of theories of how cultural ecosystems work in particular places over time. For example, Bennett (1968; also see Netting 1993) modeled the ecological dynamics of natural systems to understand sociopolitical changes in the cultural ecosystems of farmers. Cohen (1968) developed a cultural evolutionary scheme to predict settlement patterns and sociocultural development in the developing regions. Many of these cultural ecology theories have been subjected to historical critiques; nonetheless, the dynamic and predictive aspects of cultural ecosystem models are useful when examining social change on a particular site (Barlett and Chase 2004).

The case of historic Parque Central in San José, Costa Rica, illustrates this

Figure 1.3. Vendors and religious practitioners in Parque Central

Figure 1.4. Redesigned Parque Central

point. Up until 1992 Parque Central was a well-established, spatially organized cultural ecosystem made up of shoeshine men on the northeast corner (figure 1.1), pensioners on the southwest corner (figure 1.2), vendors and religious practitioners on the northwest corner (figure 1.3), and prostitutes and workmen on the center inner circle. The established cultural ecosystem, however, was disrupted in 1993 when the municipality closed the park and redesigned the historic space (figure 1.4) to remove users perceived as unattractive to tourists and the middle class (Low 2000).

The redesign, however, destroyed the social ecological balance. A new social group, a gang of young men, took over the public space, creating a dangerous and even more undesirable environment, and Nicaraguans, rather than Costa Ricans, became the main inhabitants on Sundays. This case illustrates the fragility of existing cultural ecosystems (and their diverse niches); when the sociospatial niches (places) are destroyed, the system may not be able to maintain itself any more effectively than before the intervention. In fact, the redesign of a site, ostensibly to improve it, may create more problems and dysfunction if the social ecology of the space is overlooked.

3. CULTURAL DIVERSITY

The third important dimension is cultural diversity. Biological diversity, so critical to the physical environment as a genetic repository and pool of adaptive evolutionary strategies, has its social counterpart in cultural diversity. Cultural diversity became a "politically correct" catchphrase during the 1980s in the United States, but it has not been addressed in planning and design—much less sustainable development—practice. While sustainable development includes "maintaining cultural diversity" as a conceptual goal, there is little agreement, much less research, on what it means. But cultural diversity provides a way to evaluate cultural and social sustainability, and is one observable outcome of the continuity of human groups in culturally significant places.

This modified cultural ecosystem/diversity model provides an effective theoretical basis for defining social sustainability. But social sustainability encompasses more than understanding cultural ecosystems and diversity. It implies a moral and political stance to sustain sociocultural systems—maintaining them, supporting them, and in some cases, improving them. And it is in this sense that a new series of questions must be asked. Is social sustainability applicable to all populations? We have been assuming that human ecosystems do not compete with each other, but of course they do. A successful cultural system can overrun another. Is this what we mean by sustainability—natural selection of cultural ecosystems, and the fittest survives based on an evolutionary or sociobiological model? Or should we be protecting weaker groups, systems, urban niches from stronger ones? And who is the *we*? These are moral and political questions that must be addressed in discussions of application and practice.

Ultimately, when we discuss social sustainability, we need to address issues at various scales: the local, the regional, and the global. Social sustainability at the local scale has been illustrated by the examples discussed so far, that is, understanding the cultural dynamics of a place so that specific individuals and their histories and values are sustained at or near the park or heritage site, across generations, and over time. At the regional scale, social sustainability might be better conceptualized through a broader plan that supports not only individuals but also neighborhoods, communities, churches, associations, and the institutional infrastructure necessary for the survival of cultural values and places of larger groups throughout history. Dolores Hayden's *The Power of Place* (1995; see also Hayden 1990) provides a vision of documenting and commemorating cultural histories of minorities and women that goes beyond the local and sustains larger elements of society. Social sustainability at the global scale moves closer to David Throsby's "sustainable development" based on intergenerational, and cultural, equity and environmental justice.

Thus, social sustainability is the successful maintenance of existing cultural ecosystems and cultural diversity. It is safeguarded when the systems of social relations and meanings are inclusive, rather than exclusive. In this sense, social sustainability is fostered by understanding the intimate relationship between history, values, cultural representation, and patterns of use in any culturally diverse context. In fact, the inclusion of local people, their histories, and their values ultimately strengthens any park's long-term social sustainability.

Cultural Property Rights

An equally powerful argument for cultural diversity can be made in terms of the ethics of respecting cultural property rights. At the most basic level, ethics is the consideration of the right way to live one's life, particularly with regard to interpersonal behavior (Lefkowitz 2003). But while ethics is about doing the right thing, it does not necessarily mean the same thing in each situation. Stated broadly, it is about being accountable for your actions and avoiding harm to others, but interpreted in specific social, cultural, and historical situations.

Chris Johnston and Kristal Buckley (2001), when discussing the importance of cultural inclusion in heritage conservation practice, point out that ethics translates cultural values into actions. This translation is most easily seen in cross-cultural or multicultural situations where many of the cultural assumptions and values differ. Johnston and Buckley provide the example of how the Australian Archaeological Association developed a code of ethics to regulate the principles and conduct of its members in relation to Australian Aboriginal and Torres Strait Islander peoples. "Among other things, this document acknowledges the indigenous ownership of cultural heritage knowledge and the primacy of the importance of heritage places to indigenous people" (2001, 89). In this way, the Australian Archaeological Association defined what its ethical relationship to indigenous cultural knowledge ownership would be and set boundaries for appropriate behavior with regard to indigenous peoples and their cultural heritage.

At the heart of the argument about cultural property rights are questions about who owns the past and who has the right or responsibility to preserve the cultural remains of the past. "These questions raise important philosophical issues about the past.... They also bring to the fore both the diversity of values associated with the preservation of cultural properties ... and the conflicts of interests of the various parties to the dispute" (Warren 1989, 5). Karen Warren (1989) suggests that the way to understand the various arguments that occur in a dispute is to organize them by what she calls the "3 R's": 1) the *restitution* of cultural properties to their countries of origin, 2) the *restriction* of imports and exports of cultural properties, and 3) the *retention* of rights by different parties.

Within each of these categories, numerous arguments have been used to substantiate why traditional or native cultural property rights should not be respected. For example, Warren (1989) identifies the use of "the rescue argument" against cultural property claims by countries of origin when the cultural properties at issue would have been destroyed if they had not been "rescued" by foreigners with the ability to preserve them. Those who rescued the cultural properties now argue that they have a valid claim to them. Other arguments along these lines include the "scholarly access argument"—that scholars will not have adequate access if cultural materials are returned to their country or culture of origin, the "foreign ownership argument," and the "humanity ownership argument," all of which have been used to dispute country-of-origin claims. To resolve these antagonistic disputes Warren offers an integrative perspective that emphasizes preservation as a goal and incorporates compromise and consensus models for settling cultural property matters. The importance of her solution, however, resides in her underlying ethical position that acknowledges the importance of the diversity of values and perspectives involved in any resolution of cultural heritage issues.

Museums such as the Smithsonian Institution also find themselves at the center of these ethical arguments. Ivan Karp (1992) suggests that "an acute moral dilemma is raised by the acknowledgment that museums have responsibilities to communities" (11). From this perspective questions arise about what happens when one community makes a request that hurts or constrains another community or that uses up a resource that would otherwise be shared. Museums must decide who speaks for a community and whether the claims of different groups are equally valid. In the case of the repatriation of material artifacts, local as well as national communities and cultural groups are interested in how museums make their decisions and conduct their affairs.

In order to adjudicate cultural property claims fairly, then, it is necessary that all communities and cultural groups are included in the discussion. And, we argue, there needs to be a place where they can meet and consider issues on an ongoing basis. Heritage sites and urban parks are just two examples of public spaces where these discussions can begin. The ethical imperative of cultural property rights for those whose "culture" or "environment" is being utilized or controlled by others rests on assumptions that power should be equitably distributed and that all cultural groups have rights to their native inheritance and/or home places. The same argument can be used to stress the importance of maintaining the cultural diversity of parks, beaches, and heritage sites.

Community Participation, Empowerment, and Citizenship

But cultural property rights are not the only way to think about these ethical issues. Wendy Sarkissian and Donald Perlgut (1986) give two reasons for seeking community involvement in the use of parks and heritage sites: 1) it is ethical, that is, in a democratic society, people whose lives and environments are directly affected should be consulted and involved, and 2) it is pragmatic because people must support programs and policies in order to mobilize their participation. One might add that the cost of top-down approaches to maintaining parks is staggering and that few governments can afford the economic costs of imposing external controls. Yet the benefits of collaborative approaches have not been fully realized. Even though community members who use a park often possess the knowledge and physical proximity to park resources, they are frequently not included in the planning and maintenance processes. This may be because of mistaken attitudes on the part of park administrators about the capabilities of residents and users, and because park managers do not have the staff, language, or collaborative training to work effectively with local community groups (Borrini-Feyerabend 1997).

Discussions of community participation and empowerment have become increasingly important as cities have become more ethnically diverse and more demographically and racially divided (Gantt 1993). Parks that originally served relatively homogeneous white middle-class or working-class neighborhoods must now provide recreation, educational and social programs, and relaxation for an increasingly multicultural and multiclass population. Mayors and city council members, as well as park managers and planners, are hard-pressed to mediate the conflicts that arise as park resources are stretched thin and as neighborhoods deteriorate because of the inability of local government to provide adequate services for all residents. And as we already know from the history of decreasing municipal funding, parks and heritage sites are low priorities when education and health care needs loom large.

The question arises, then, whether increased cultural diversity in the city can be utilized to improve the lives of residents (Gantt 1993). We argue that it can by empowering local groups to voice their needs and claim their histories in both local and national park contexts. By empowering communities to claim park resources as their own and to engage in the decision-making process that allocates funds and labor for park maintenance and programming, park managers gain collaborators in keeping the park well-attended, safe, and well-maintained. At the same time, city administrators and park planners learn more about the diverse needs of ever-changing neighborhood social and cultural groups and their values, making it possible to more accurately match cultural group needs with available resources.

There are a number of urban programs that have used community participation and empowerment strategies to structure the running of local cultural resources and park offices. For example, the "Charleston Principles" of Seattle, Washington, require that any proposed change include a community cultural planning process involving a broad spectrum of community members—public agencies, civic and social groups, educators and students, business and economic interests, artists, community leaders, and cultural organizations of all types. In this way, community empowerment is a legally mandated part of any planning and design process (King County Landmarks and Heritage Program 1999).

Another example is "Taking Action," a project in Australia that has produced a handbook for actively involving communities in heritage projects (Johnston and Clarke 2001). Using the same ethical and practical arguments we have discussed here, the authors see community involvement as part of participatory democracy whether a project is run by an elected government or initiated and directed by the community itself. By involving the community, it is possible to: 1) understand community aspirations and values, 2) find out about community needs, 3) learn about the locality and community, 4) share perspectives, 5) find out about differences as well as similarities, and 6) ultimately create new solutions that draw upon a wider range of ideas (2001, 3). Johnston and Clarke's report supplies a checklist of ways to communicate with people and involve cultural groups, and it is an excellent guide for beginning any community involvement project.

Other collaborative programs emphasize the inclusion of indigenous communities often overlooked in park planning and administration and marginalized by local politics. Barbara Harrison (2001) summarizes the experiences of working with indigenous groups and researchers in North America as well as New Zealand and Australia to develop her guide to collaborative working relationships in research and applied practice.

The concept of citizenship, and its accompanying rights, underlies each of these projects. The liberal notion of citizenship defines people as individuals who have civil, political, and social rights within the nation-state. But this definition is limited in that citizenship must also be considered full membership of a community within a neighborhood, region, or state, and membership of individuals within one or more community groups. Citizenship should be understood as inclusive of state, regional, neighborhood, and community levels of individual participation, thus producing a multistranded and multilayered model of the sociopolitical relationship of people and society (Yuval-Davis 1998).

Most debates over citizenship are about the basic right of entry into a country—whether a person can stay, maintain a residence, and not be repatriated—and about work-status issues, participatory duties such as voting, and availability of social welfare benefits. But these same notions can be applied to the rights of individuals and groups to participate in decisions about places, resources, and services that touch their lives. We argue that citizenship also should focus on the role that individuals and communities play in determining the success or failure of their local open spaces and historic resources. Full citizenship includes community involvement and participation in the ongoing life of the neighborhood and region, and as such it provides another justification for community empowerment and participation in park planning processes. If all community and cultural groups are included, then we are also empowering citizen-leaders and participants who will continue to contribute to the area and its growth and stability over time.

Dissonant Heritage, Negative Heritage, and the Politics of Meaning

With the empowerment of community and cultural groups, however, there emerges a set of problems and conflicts that J. E. Tunbridge and G. J. Ashworth (1996) have called "dissonant heritage." The concept of dissonant heritage is derived from the idea that heritage is a contemporary product shaped by history in which different narratives exist. Dissonance in heritage suggests a discordance in these histories and lack of agreement and consistency in the way that the past is represented (Tunbridge and Ashworth 1996). Dissonant heritage is present whenever there is more than one meaning for an object, place, or landscape; most often it is embedded in a conflict between tourism and sacred use of a site or between global and local meanings (Graham, Ashworth, and Tunbridge 2000).

The creation of any heritage site—and any park, we would add—"potentially disinherits or excludes those who do not subscribe to, or are embraced within, the terms of meaning defining that heritage" (Graham, Ashworth, and Tunbridge 2000, 24). It is a common condition in multicultural societies in which inclusiveness is determined by a group's proximity to political and economic power. Despite the development of pluralist societies, heritage—and many other aspects of the landscape and built environment—often reflects only the dominant culture. Certain European societies typically do not acknowledge their former colonial subjects (Graham, Ashworth, and Tunbridge 2000), while white Americans often avoid recognizing their being the beneficiaries of slavery and the early dependence on slave labor in the plantation economy.

Kenneth E. Foote (1997) addresses these issues of unresolved meaning and the politics of memory by arguing that the invisibility of some violent or tragic events, especially those dealing with minority populations such as African Americans or Latinos, indicates a certain tolerance or acceptance of such events as part of American life (294). Other tragic events, such as the Battle of Gettysburg, are celebrated as fundamental to understanding the American past. This dual tendency—to ignore and to celebrate—reflects Americans' ambivalence toward events that both bind and divide us and "casts un unusual shadow over American history and the American landscape" (Foote 1997, 294). Thus, the practice of telling all sides of the story and of uncovering uncomfortable and conflicting views of the past that produce dissonant heritage has never been popular. But the pervasiveness of dissonant heritage is vital to our discussion of urban parks and public spaces in that it provides another rationale for why cultural diversity and community inclusiveness are so important. The negotiation of dissonant meanings and their resolution in forms representative of all cultural groups and communities is the ideal toward which we should be working.

Cultural Values

In historical preservation practice "values," like ethics, means the morals and ideas that guide action as well as the specific qualities and positive characteristics of things as seen by a particular person or group (Mason 2002). Sociological approaches consider values "generalized beliefs about what is or is not desirable, but also as motives . . . that influence people's actions" (Feather 1992, 111). Psychologists such as Joel Lefkowitz (2003), on the other hand, define values as "relatively stable cognitive representations of what the person believes are desirable standards of conduct or generalized end states" (139; also see 151); Lefkowitz adds that values have emotional and evaluative importance to one's ideal self-concept, and provide motivation for people's actions and choices. In our discussion, we draw upon elements of each of these definitions and utilize the concept to refer to the meanings and feelings, positive or negative, that people attribute to their lives, environment, actions and behaviors, and world as a whole. Values, however, are not inherent in an object, action, or landscape but are contingent on the circumstances—the place, time, and company—in which a judgment is being made. As opposed to the psychological definition of values as relatively fixed and stable within a person, our perspective identifies community values as often fluid and changing, although they may be relatively fixed depending on the domain.

"Cultural values" refers to the shared meanings associated with people's lives, environments, and actions that draw upon cultural affiliation and living together. They are often expressed as value judgments, that is to say, something is considered bad or good depending on how it registers with a person's or group's attitudes at a particular moment. These value judgments, usually expressed as liking or disliking some person, place, or object, provide information about underlying unspoken cultural assumptions, beliefs, and practices. Cultural values are our best indicators as to what people think and feel about a landscape such as a park or heritage site, and they can act as a guide to understanding park use and disuse, place attachment or lack of it, and symbolic meanings. According to Randall Mason, "sociocultural values are at the traditional core of conservation—values attached to an object, building, or place because it holds meaning for people or social groups due to its age, beauty, artistry, or association with a significant person or event or (otherwise) contributes to processes of cultural affiliation" (2002, 11).

We would add that cultural values also accrue to objects, buildings, and landscapes through living in a place for a long period of time, working in a place, narrating stories and telling myths about a place, and engaging in any activity that would generate a relationship between a person or group and a particular location. This kind of "cultural place attachment" (Altman and Low 1992; Low 1992) often develops between people and places, particularly places such as parks, beaches, and heritage sites that have potential meaning and cultural significance through their ongoing use and role in memory making.

One important concern when discussing cultural values is that the term *cultural* is politically as well as socially constructed and manipulated for a variety of ends. Cultural values, similar to cultural identities, are not necessarily definable attributes that can be measured or codified, but they must be understood as negotiated, fluid, and context-dependent. The political importance of a neighborhood can change depending on how the residents present themselves and their values to the various players involved. Sociopolitically constructed cultural labels such as black, African American, white, Jamaican, or Haitian evoke different meanings and responses from New York City officials and planners and are actively manipulated by the community in neighborhood descriptions and media coverage (Low 1994). Poor people and their values, however, are often the most vulnerable because the local constituency does not have the political and economic power to struggle against the definitions and decisions of government officials and private entrepreneurs.

Further, processes of cultural hegemony—that is, the preeminence of one cultural group's ideas and values over another's—maintain the control of middle- to upper-middle-class white values over the definitions of what can be considered relevant to other cultural groups in a neighborhood or region (Lawrence and Low 1990). The values of planners, managers, administrators, designers, and National Park Service employees are also hegemonic because of the entrenched belief that professionals know more than the local community. Yet when elites and professionals dictate what should happen to an urban space, their landscape preferences do not necessarily correspond to the needs and desires of the local users.

Cultural values and their representation in park planning and renovation processes are decisive in producing programs that will work in a specific community location. Prospect Park, discussed in Chapter 3, is an excellent example of how local cultural values do not necessarily match the values of the professionals who are managing the park and making decisions about renovations and financial investment in the park's future. Relying on professional expertise rather than taking seriously cultural values about park resources reinforces the traditional inequality of power relations and exacerbates race and class conflict already in evidence. Another example of the importance of understanding cultural values is discussed in Chapter 4, on the Ellis Island Bridge Proposal. Historic preservationists did not understand why it would be important to build a bridge for local residents until they confronted the value placed on visiting the park in large family groups by the black community. Suddenly the $7.50 price of a ferry became $75.00 for 10 family members, putting visiting or attending programs or activities out of the reach of these families.

What Is Cultural Diversity Good For?

Ulf Hannerz (1996) suggests that the value of diversity is so entrenched in the contemporary discourse about culture that it is difficult to reflect clearly on it. So he offers what he calls his "seven arguments for diversity" to make the point that there are many basic reasons to consider cultural diversity important to our lives. He includes many of the points that we have made in this discussion and adds others that we have not emphasized, arguing that cultural diversity is important because it provides:

1. the moral right to one's culture, including one's cultural heritage and cultural identity;
2. the ecological advantage of different orientations and adaptations to limited environmental resources;

3   a form of cultural resistance to political and economic domination by elites and power asymmetries and a way to counteract relations of dependency;
4   the aesthetic sense and pleasurable experience of different worldviews, ways of thinking, and of other cultures in their own right;
5   the possibility of confrontation between cultures that can generate new cultural processes;
6   a source of creativity; and
7   a fund of tested knowledge about ways of going about things. (Hannerz 1996, 56–57)

We would add that attention to cultural diversity also leads to community empowerment, expanded citizenship, and the involvement of people in the governance and maintenance of their neighborhoods and workplaces. It expands the notion of individual rights of citizenship to include the survival of one's culture and/or cultural group, and the marking of its importance in the landscape. We would also add that creativity from cultural contact and interaction flows from cooperation as well as from working out solutions to conflicts and confrontation. Therefore, cultural diversity, utilized effectively and honestly, leads to more democratic practices and peaceful relationships between people within a locality especially if all groups are treated equally with respect for their needs, desires, and adequate space and resources for work, home, and recreation.

We end this introduction where we began, by asserting how crucial understanding cultural diversity and community values is to having a successful park, beach, or heritage site. Assessing social and cultural values remains the best way to monitor changes in the local neighborhood or region, and we offer a number of ways to elicit and collect these values in the following examples. Each case study emphasizes one of the lessons for large urban spaces. For example, Independence National Historical Park focuses on cultural representation and its impact on local group attendance. But each case also encompasses all of the lessons. Any inclusive urban space exemplifies many of these principles and others that we have not yet examined.

This book begins a conversation between social scientists—anthropologists and environmental psychologists—and the decision makers who direct, design, plan, and manage our nation's parks, beaches, and heritage sites. The goal is to contribute what we have learned from our research experiences to making urban parks the best places they can be for the most people. Parks offer urban residents a place away from home that is essential to their physical and mental health and well-being. This is particularly true for the poor and working-class residents who do not have backyards, much less vacation homes, where they can rest and recreate. We hope the lessons and the research on which they are based help to improve and promote these socially important and wonderful places—the urban parks, beaches, and heritage sites of New York and the rest of the Northeast.

Organization of This Volume

The book includes case studies drawn from our research on National Park Service parks, seashores, and heritage sites: The Ellis Island Bridge Proposal (Chapter 4), Jacob Riis Park in the Gateway National Recreation Area (Chapter 5), and Independence National Historical Park (Chapter 7), as well as two case examples drawn from our work on New York City parks: Prospect Park (Chapter 3) and Orchard Beach in Pelham Bay Park (Chapter 6). Chapter 8 provides the methodological background and specific anthropological research techniques used to gather these data for those interested in undertaking this type of research in their own parks and communities. The conclusion revisits the six lessons we identify for promoting, maintaining, and managing cultural diversity in urban parks and reflects on what was learned from this long-term research project on urban park policy.

Chapter 2
Urban Parks
History and Social Context

As Michael Brill (1989), Sam Bass Warner (1993), and perhaps others have noted, the variety of park types has multiplied since parks first appeared in North America in the early nineteenth century. Many kinds of public spaces fall under the general rubric of "park." The case studies in this volume are a sampling of urban park types: a landscape park, two recreational beach parks, and two historical parks. To situate these cases from New York and Philadelphia within a national context, this chapter provides a comparative review of the history of various park types in the United States.

The first urban parks in the United States were relatively unimproved commons, places originally set aside for grazing cattle and training militias. New York's original common is now the heavily gated City Hall Park. Boston Common is perhaps the best example of the type. Set aside only six years after the original settlement, Boston Common has maintained its 44 acres and something of the informal, unornamented character of a colonial common. Straight, paved paths lined with benches crisscross its territory in practical fashion, enabling people to cross over easily in their travels about town. Large trees shade the grass-covered ground with no shrubs, ornamental trees, flower beds, or other plant varieties to complicate the picture. The Common has several frankly recreational facilities: tennis courts, ball field, children's playground, and seasonal skating/wading pond. Like many smaller city squares in New York and elsewhere, Boston Common is more an extension of urban space than a refuge from it. No perimeter plantings screen the surrounding cityscape from view. Rather, much of the character of the place comes from the visibility of adjacent structures from within the grounds.

Boston Common was less parklike before the early nineteenth century. The 1820–1840 period brought a movement to create tree-lined paths for strolling by the fashionable citizens who lived nearby (Domosh 1998). The formal paths and tree-lined promenades date from this period, and the practice of grazing cattle was ended. Similar improvements were made at this time to the coarse open spaces of town commons and squares throughout New England, New York, Pennsylvania, and Ohio. Philadelphia's original five squares were similarly devoid of landscaping until this time, when paths were laid and trees planted. Today, Rittenhouse Square functions like Boston Common as a simply planted, central open space that complements the urban concentration that surrounds it. Full of people sitting and strolling, lying on the grass, playing ball, or listening to buskers and soapbox orators, these simple places are about as close as eastern North American cities come to the Latin American plaza. J. B. Jackson (1984) stresses the essentially political character of these urban plazas: in them one is revealed as a citizen.

The Landscape Park

Urban landscape parks, beginning with Central Park in New York, have quite different origins. Typically much larger than squares and commons, they were designed as refuges from the city according to an exacting aesthetic formula that simulated the idealized English and North American countryside. Prospect Park in Brooklyn, among the best examples of the type, encompassed 526 acres and incorporated pastures, woods, gathering places, and systems of surface waters, carriage drives, and footpaths. It was designed by the firm of Frederick Law Olmsted and Calvert Vaux, beginning in 1866, several years after their first and most famous design collaboration that produced Central Park. Unlike the older urban squares, Prospect Park kept the surrounding city out of view with a high, thickly planted earthen berm along its perimeter.

Prospect Park was a product of the park movement that swept through North America during a 50-year period beginning in the 1840s. The movement had philosophical, theological, and nationalistic sources. The philosophical basis lay in romanticism and its belief that nature and natural scenery had the power to uplift and restore the human spirit. Romanticism arose in reaction to the effects of industrial capitalism evident already in the 1840s and 1850s—rapidly growing cities, tenement housing crowded with immigrants, factory life, epidemic disease, and smoke. Romanticism took many forms of expression, one of them being landscape gardening. The landscape gardener sought to arrange nature's best qualities in prospects of quiet repose. The romantic sensibility in gardening called for a naturalistic imitation of nature, rejecting the once dominant baroque design idiom of straight lines in formal perspective.

The new parks had several precedents, including new public parks in England, the older royal parks in many European cities that had by then been opened to public use, and rural cemeteries like Mount Auburn in Cambridge, Massachusetts. Mount Auburn's designers strove for "a picturesque effect" composed of serpentine walks and paths, groves of dark woods, ponds, clearings, and ornamental plantings of trees, shrubs, and flowers (Von Hoffman 1994, 73). The garden cemetery idea soon spread, bringing about Green-Wood Cem-

HISTORY AND SOCIAL CONTEXT

Figure 2.1. Romantic detail—Cleftridge Span in Prospect Park

etery in Brooklyn, Laurel Hill Cemetery in Philadelphia, and others. Mount Auburn Cemetery and its progeny soon became popular resorts for outings and picnics among middle-class city dwellers. The rural cemetery was an important precursor to the urban landscape park in demonstrating the popularity of a romantic landscape of winding paths, groves of trees, ponds, and beautiful views. These cemeteries whetted the public appetite for large parks.

Contrasting vernacular traditions in recreational landscapes coexisted with the development of formal parks. One such landscape tradition was the undesigned and unplanned, but popular, common open space. In the small town and growing city alike, informal open spaces lying just outside the developed area were appropriated for outings, get-togethers, picnics, sports, and games. These spaces are hard to document because they were not formally planned, designated, or designed, and most gave way to urban development long ago. Jackson (1984) contrasts the formal town park of the mid-nineteenth century—

typically very pretty but empty of people—with the lively grove just outside town, often along a river. There, on level ground under the big cottonwood trees and along grassy banks, townspeople would gather on a Sunday afternoon for informal activities of all kinds. Such places outside the larger cities were largely working-class resorts avoided by more fastidious citizens: "they were crowded, boisterous, and sometimes violent" (Jackson 1984, 114). In sharp contrast, the new landscape parks substituted the aristocratic garden for the oldest, most popular kind of play space—sizable areas where common people could exercise and play and enjoy themselves and participate in community life (Jackson 1984).

Another vernacular tradition was that of the commercial pleasure ground. Several popular pleasure grounds existed in and around New York: Rosenzweig and Blackmar (1992) cite Niblos, Palace Gardens, at Sixth Avenue and 14th Street, and Harlem Gardens. Jones Wood, around 61st Street along the East River, was considered as a site for what became Central Park. Hoboken, New Jersey, offered the Elysian fields, one of the places where baseball was first played, which included level open ground and a landscaped eminence overlooking the Hudson River. Another popular resort for day trips stood at New Brighton, Staten Island. London had its pleasure grounds, among them Vauxhall Gardens, Ranelagh, and Cremorne Gardens (Whitaker and Browne 1971). One of the few such places that survives today is Copenhagen's Tivoli Gardens.

The pleasure grounds "liberally mixed all styles of art and decoration to create recreational spaces that responded to popular desires for novelty and diversion." Their eclectic style featured statues, fountains, grottos, arbors, artistic displays, and tents for refreshments or performances. "Lively crowds engaged in picnics, festivals, and sports in the shady groves and open pastures of former farms or gentlemen's country seats" (Rosenzweig and Blackmar 1992, 104).

During the planning of Central Park, some New Yorkers hoped for a synthesis of the vernacular aesthetic of the pleasure ground with the English naturalistic landscape tradition exemplified by the rural cemetery.[1] Most park proponents, however, regarded the eclectic enticements of commercial pleasure grounds as vulgarities. Whether Central Park would adopt the English landscape style or the characteristically geometric style of French and German parks, the influence of gentlemen gardeners and other sophisticated advocates meant that Central Park was certain to adhere to a strict aesthetic standard.

Although planned as places of healthful recreation for all classes, landscape parks were built to middle-class standards (Taylor 1999). This environment of seamless coherence between polite middle-class behavior and a graceful, tastefully furnished landscape would "naturally" compel the working-class users to emulate their social betters. If emulation was not forthcoming, widespread

HISTORY AND SOCIAL CONTEXT

supervision and enforcement effectively curtailed unsuitable behavior. In addition to active sports, working-class recreation in the nineteenth century often involved excessive drinking, exuberant park play, demonstrations of power, and loud, rowdy behavior, in compensation for the rigor, monotony, and boredom of the job. The conflicts that sometimes occurred between working-class behavior and middle-class mores resulted in the criminalization of certain forms of behavior (Taylor 1999).

The Movement Spreads

Other parks in North America designed by Olmsted and Vaux or Olmsted and other collaborators include Franklin Park and the Arnold Arboretum in Boston; Delaware Park in Buffalo; Highland and Seneca parks in Rochester, New York; Lake Park in Milwaukee; Cherokee, Iroquois, and Shawnee parks in Louisville, Kentucky; and Mount Royal Park in Montreal. Olmsted designed a great many other landscapes as well—residential subdivisions and one entire suburb (Riverside, Ill.), campuses, cemeteries, and private estates including Biltmore, near Asheville, North Carolina. Many more city parks designed by others have much in common with the Olmsted tradition. Some were the work of Olmsted's sons and successors, the Olmsted Brothers.

Rather than preserving existing landscapes of high scenic and ecological value, like so many later park projects, these early parks were designed and built often on degraded sites. Olmsted and others of the time wanted to create great social spaces out of the materials of nature. The lakes, streams, waterfalls, and pastures were created. Part of the great expense of building Central Park, for example, was the cost of blasting rock outcrops to create the level expanse of the "Sheep Meadow." Ample provisions for public use were built too: the carriage drives, concert groves, promenades, "refectories," and boathouses.

Olmsted himself fought to keep the balance in his parks in favor of natural surfaces, warding off demands to give space over for recreational facilities, for museums and zoos, and for monuments and memorials. Heckscher (1977) shows how Forest Park in St. Louis filled up over time with such facilities. Designed by Olmsted followers in a restful Olmstedian configuration of curvilinear drives, open fields, and groves of trees, the park was transformed by serving as the site for the St. Louis Exposition of 1904; many trees were felled, leaving exhibition buildings in place. "Thereafter, every time a new institution was created or a new entertainment devised, Forest Park seemed the natural place to put it. Today the park contains a zoo . . . , a fine arts museum, a planetarium, an ice-skating rink, a municipal opera and three golf courses, as well as much space given over to parking" (Heckscher 1977, 176). In recent years, Forest Park has undergone a comprehensive regeneration program aimed in part at restoring some sense of the natural amid its collection of civic institutions and recreational facilities (*Landscape Architecture* 1998).

Public Reservations and State Parks

Pelham Bay Park, the other New York landscape park studied in this volume, is representative of the paradigm shift in metropolitan parks after the first run of landscape parks. This was the metropolitan woodland reservation, pioneered in the United States during the 1890s in the Boston area, under the leadership of Charles Eliot, a partner in the Olmsted firm. Olmsted's work for Boston had already produced a chain of constructed parks within the city limits linked by landscaped carriage drives, or "parkways." Olmsted's Boston work modeled his belief that cities needed not just a single park, but a park *system* to bring the benefits of natural scenery within walking distance of all residents. Eliot and others built on this idea by working to create a system of public grounds over the whole metropolitan area. In its land acquisitions, the Metropolitan Park Commission focused on areas of intrinsic scenic value to create large "reservations" of wooded and watered land open to the public. These they linked to the urban core with a series of parkways and boulevards—still motorless in the 1890s. A second focus of the commission's acquisitions was beaches and riverbanks. Much of the work of establishing public reservations in shoreline areas required the acquisition and clearance of private structures and the building of parkways along them to provide access (Haglund 2003).

Metropolitan reservations like Boston's, or the larger Chicago Forest Preserve system, were different from the earlier parks in preserving existing landscapes rather than creating idealized scenery. Eliot urged minor modifications such as cutting trees that blocked scenic views over a valley or maintaining existing clearings for their scenic value. Streetcar companies built lines that brought the reservations within reach of city dwellers lacking their own carriages. Still, these were mainly unimproved places. One assumes they had something of the feeling of the wooded "grove" that attracted informal recreation and sociable gatherings (Jackson 1984).

The idea of reserving scenic land from development through public acquisition led to the creation of state and county parks all over the country. Today every state has a system of public recreational lands, usually in areas with recreational potential, having one or more elements such as valuable forests, mountainous or rugged terrain, and attractive surface waters. Such parks invite hiking, cookouts and picnics, swimming, boating, and fishing; some permit overnight camping.

The state park movement began when California established a state park in the Yosemite Valley and the Mariposa Big Tree Grove in 1866 (later Yosemite National Park). The state acted in part from a report submitted by Frederick Law Olmsted, which urged that such magnificent scenery be made public to uplift the health and spirit of the large majority of citizens who would otherwise not have access to such natural resources (Newton 1971).

Yosemite would remain the only state park in the country until New York created the Niagara Falls Reservation in 1885. That same year, in an atmosphere of inflamed public opinion over logging excesses in the Adirondacks, the New York legislature established the Adirondack Forest Preserve to protect the remaining state-owned land in the Adirondack region from being sold off to loggers. The act provided that lands within the preserve would be "forever kept as wild forest lands" (Terrie 1994, 92). Although the land—and therefore, it was believed, the headwaters of the state's major rivers—would be protected, it was assumed that the timber would be managed according to scientific forestry principles. An 1895 constitutional convention, however, took the then-radical step of adding a provision that protected the timber in the preserve from being "sold, removed, or destroyed" (Terrie 1994, 104).

Other early state parks include Lake Itasca State Park, containing the headwaters of the Mississippi River in Minnesota, in 1891, and the first county park organization, in Essex County, New Jersey, in 1895. Illinois established the Starved Rock State Park in 1911; Wisconsin formed a State Park Board in 1907, and Connecticut founded its State Park Commission in 1912 (Newton 1971). In general, the impetus for state parks concerned public enjoyment and recreation rather than averting environmental catastrophe. The state park movement took its biggest strides in the 1920s and 1930s, with a strong impetus from Stephen Mather, director of the National Park Service. By that time the National Park System already comprised many of its most famous units: Yellowstone, Yosemite, Glacier, Grand Canyon, and Crater Lake, among others. The public interest in touring parks amid the spectacular growth of automobile ownership led to grave concern about overuse of national parks for recreational purposes. It was hoped that a system of well-distributed state parks would satisfy much of the desire for recreation and act to buffer the national parks from overuse (Newton 1971; Cutler 1985).

Pelham Bay Park in the Bronx belongs more to this era of public reservations than to the earlier era of constructed parks. Selected for the scenic value of its shoreline and adjacent islands along Long Island Sound, the park incorporated the estates of several wealthy families. One of these, the Bartow-Pell Mansion, has been preserved with its gardens and grounds. At 2,700 acres, Pelham Bay Park is the largest unit in the New York City park system.

The park's most popular attraction is a crescent-shaped saltwater beach with a promenade and extensive recreational facilities opposite the beach. These—including the beach itself—were constructed by Robert Moses's Parks Department during the Depression years using federal funds available under the New Deal. The beach and its adjacent facilities occupy an area of landfill that connected the two largest offshore islands to the mainland. The park includes woods and wetland tracts, two golf courses, picnic grounds, and riding stables.

Van Cortlandt Park, in the north-central Bronx, has a similar history. The park preserves rocky woodland and river shoreline as well as the building and grounds of the colonial-period Van Cortlandt estate. Within its 1,146 acres, Van Cortlandt Park offers the nation's oldest municipal public golf course, nature trails, athletic fields, swimming pool, and athletic stadium. Both these Bronx landscape parks were created in the late nineteenth century out of country estates lying at the far reaches of the territory newly annexed by New York City from Westchester County. The two Bronx parks were part of a comprehensive city planning effort that mapped streets and parkways in the new borough of the Bronx in anticipation of widespread population growth. In its first 40 years Pelham Bay Park was shaped as much by vernacular place-making activity by users as it was by the Parks Department. As described in Chapter 6, the park has a fascinating history of tent colonies and garden-building groups. Not until the WPA years of the mid-1930s did the city government develop the extensive picnic grounds and athletic and swimming facilities of Orchard Beach.

The Recreation Facility Park

The 1920s and 1930s represented the crest of a third era in municipal parks, emphasizing recreation facilities, which began at the turn of the century. The recreation facility park recalls the Olmsted park in having been largely constructed. The atmosphere, however, was entirely different from either the landscape park or the public reservation. In both landscape parks and woodland reservations, providing users contact with nature had priority over active forms of recreation. The recreation facility park had its roots in the Progressive movement at the turn of the twentieth century. The reformers of that era believed that park planners needed to take an activist stance in bringing the benefits of wholesome recreation to urban people, especially children. This goal was realized in the playground, a facility provided with specialized recreational spaces and equipment and staffed by play directors. Olmsted Sr. had designed an "outdoor gymnasium" in the West End of Boston in 1892, but landscape architects regard Chicago as the standard bearer in the playground movement (Newton 1971; Cranz 1982). The Olmsted Brothers' designs for neighborhood parks in Chicago's South Park District became models for the relatively small, rectangular park that provided structured recreational facilities under professional supervision.

Construction of recreation facility parks increased exponentially in the 1930s once the federal government began funding local park construction. Under Robert Moses's leadership, New York far exceeded any other city in acquiring federal funds for park construction. Standardized playground construction flourished, as did swimming pools and beachside parks. Orchard Beach within Pelham Bay Park is a good example of the Moses approach. The beach itself was entirely constructed by means of filling in the intertidal area to connect two former islands with a mainland peninsula. The park featured a crescent-shaped beach and concrete promenade, the beach sand having been hauled in from 40 miles away. The promenade gave access to handball and basketball courts. Two spacious buildings—a "gleaming restaurant" and a bathhouse—stood at the center point of the promenade (Cutler 1985). A divided parkway led visitors to the beach through the woods of Pelham Bay Park, arriving at a giant parking lot. A landscaped mall then led from the lot toward the beach and bathhouse. At either end of the beach and in back of the handball/basketball courts, existing wooded slopes were modified to create extensive picnic groves with tables and grills. Here picnicking families could look out from a height toward the beach and Long Island Sound.

Jacob Riis Park is a product of the same era. Located on a stretch of oceanfront beach along the Rockaway barrier-island peninsula in Queens, Riis Park was remodeled and much enlarged under the New Deal park-building regime directed by Robert Moses. Moses's landmark career as a political power broker in mid-twentieth-century New York has been acidly described by Robert Caro (1974). Moses had personal ambition, a gift for writing legislation, and an interest in large-scale urban planning. Taking advantage of the opportunities for construction brought first by the New Deal and later, after World War II, by federal highway and housing programs, Moses became a kind of planning and construction czar over New York City and Long Island. His record includes public housing projects, parkways and interstate highways, six major bridges and tunnels across the local waters, hundreds of park and playground projects, and the World's Fairs of 1939–40 and 1964–65.

From the 1930s through the 1950s, the locations and designs of parks around New York were in many cases the personal choices of Robert Moses. Moses loved the salt water and enjoyed swimming in it. His attachment to the bays and barrier islands of Long Island inspired him to create a system of landscaped parkways and sumptuous waterfront parks facing ocean and bay. The design and layout of Riis Park and Orchard Beach owe a debt to Moses's first and greatest beach project, Jones Beach, located on a barrier island near Freeport, Long Island, which opened in 1929.

Prior to Jones Beach, a bathhouse meant a functional pavilion of wood or concrete construction. A water tower was a tank mounted atop stilts. Public beaches were customarily lined with snack bars and other piecemeal commercial attractions and amusements. At Jones Beach, Moses worked out the new idea of combining a beach with a spacious, elegantly appointed park. The central water tower at Jones was designed to resemble the campanile of St. Mark's Cathedral in Venice. The two bathhouses contained restaurants and snack bars, rooftop terraces, showers and changing rooms, and giant swimming pools. Each was finished with expensive materials and elaborate architectural details in Moorish and Art Deco styles, all in a setting of oceanfront promenades, landscaped walkways, game areas, and lawns and picnic groves (Newton 1971; Caro 1974). One arrived at Jones Beach over a system of Long Island parkways also constructed by Moses, including one that ran on the bay side down the length of the barrier island.

Neither Riis Park nor Orchard Beach were built on the scale of Jones Beach, but both were designed by Moses's chosen architects not many years after Jones Beach. Riis Park, in particular, resembled Jones in its elaborate bathhouse, landscaped walks, ample playgrounds, and game areas, all facing a wide ocean beach. All three parks were designed to celebrate arrival by automobile and thus provided massive parking lots.

National Parks and Heritage Sites

Within the context of this book, Riis Park provides the logical transition from municipal and state parks to a discussion of the National Park System. Originally a municipal park, Riis Park became part of the national system in 1973 with the creation of Gateway National Recreation Area.

Gateway is a relatively new form of national park as it combines many different kinds of spaces. Most of Gateway's 26,000 acres comprise the islands, wetlands, and water in Jamaica Bay, the westernmost of the intercoastal bays that separate the south side of Long Island from its chain of barrier islands. Lying entirely within New York's city limits, Jamaica Bay is a fairly degraded body of water. Improperly sited municipal landfills along its margins leach contaminants into the waters of the bay, much of its naturally cleansing wetland shoreline has long since been preempted for urban development—including Kennedy International Airport—and contamination keeps all the bay's mudflats legally closed to shellfish harvesting. Still, as a surviving, large-scale estuary in an intensively urban region, Jamaica Bay is an important ecological resource. Gateway was created in part to preserve the bay as a wildlife habitat, and one of its more significant features is the Jamaica Bay Wildlife Refuge.

Gateway includes considerable territory on two barrier-island peninsulas, Rockaway in Queens and Sandy Hook in New Jersey. These peninsulas were the sites of military installations, Fort Tilden in Queens and Fort Hancock at Sandy Hook. Jacob Riis Park is located on the Rockaway strand immediately east of Fort Tilden. Gateway includes two other surplus military facilities: the former Floyd Bennett Field Naval Air Station, located on filled land alongside Jamaica Bay, and Fort Wagner on Staten Island, one of two nineteenth-century batteries that guard the approach to New York Harbor from the heights above the Narrows.

National recreation areas like Gateway illustrate the degree to which the mission of the National Park System has broadened since the inception of the system's first unit, Yellowstone National Park, in 1872. The mission has been not only to provide for public enjoyment but also to conserve natural and historic resources. National parks are created to serve three purposes, in varying degrees: 1) scenic values, 2) scientific values, and 3) historical values (Newton 1971), but the interpretation of these values has changed over the years. The importance of scenery is obvious in the selection of characteristic national parks like Yosemite; less so is science, though Yellowstone, Crater Lake, Mount Rainier, and others contain natural phenomena of great importance to scientific inquiry.

The preservation of sites of primarily historic or prehistoric importance began with Casa Grande in Arizona, designated a "ruin reservation" in 1892. Feeling had been brewing for many years that forceful measures should be taken to protect Native American sites in the Southwest from grave robbers and other desecrators. These concerns led to the passage of the Antiquities Act of 1906, which gave presidents authority to reserve "historic landmarks, historic and prehistoric structures, and other objects of historic or scientific interest" on federal lands as "national monuments" (Mackintosh 1991, 13). The Antiquities Act was used to preserve many such prehistoric sites—although the best-known site, Mesa Verde in Colorado, was separately established as a national park by Congress in the same year, 1906.

The National Park System had no dedicated management service until 1916, relying until then on U.S. Army deployments. The lack of a dedicated park service led to the first galvanizing environmental crisis of the twentieth century: congressional approval in 1913 of San Francisco's petition to dam the Hetch Hetchy Valley in Yosemite National Park for a municipal water source. The "rape" of the Hetch Hetchy had been bitterly fought by prominent Californians, including John Muir, a founder of the Sierra Club. With that battle lost, activists, including Stephen Mather, a Chicago businessman, worked to establish a federal park service. President Wilson signed legislation in 1916 to establish the National Park Service within the Interior Department, naming Mather the first director (Mackintosh 1991).

There were few national parks in the eastern United States prior to 1933, mainly because the federal government owned little land east of the Mississippi. Two early eastern parks, Acadia and Great Smoky Mountains, were created largely by donations from private landowners. The Service's second director, Horace Albright, sought to increase the system's presence in the East by acquiring historic military sites, until then managed by the War Department. Albright convinced the newly inaugurated President Roosevelt to adopt this approach, and in a 1933 reorganization a host of battlefields, monuments, and historic sites were reassigned to the National Park Service (Mackintosh 1991). Included within this reorganization was the Statue of Liberty National Monument.

Among the early urban park projects of the National Park Service, as authorized by the Historic Sites Act of 1935, were the "National Expansion Memorial," in St. Louis, Missouri, in 1935, and the Salem Maritime National Historic Site, in 1938. Work at the St. Louis site culminated not in preservation at all, but rather in site clearance and construction, in the 1960s, of Eero Saarinen's Gateway Arch. In 1948 Congress authorized another major historical project in an urban setting, Independence National Historical Park. As explained in Chapter 7, Independence did preserve important sites and objects associated with the Declaration of Independence, although it too involved extensive demolition of old buildings not then deemed to have historic value.

Eager to extend its domain further in the populous East, the NPS advocated establishing national seashores and even urban recreational parks in the 1930s. Only one national seashore, Cape Hatteras, was established before World War II. Cape Cod National Seashore entered the system in 1961, followed by Point Reyes, California, in 1962; Fire Island, New York, in 1964; and Indiana Dunes National Lakeshore, near Chicago, in 1966 (Foresta 1984).

The proximity of most of these seashores to metropolitan areas fit in with the Great Society's mission of evening out the unfair distribution of public goods (Foresta 1984). The procession of national seashores was also a response to the rapid suburban growth of the postwar era. National seashores would prevent overdevelopment of unique natural resources within their boundaries. National seashores were still some distance from the densest population centers, many of which were at the time places of increasing unrest. Plans were advanced in the late 1960s and 1970s for a new form of national park, the "national recreation area." Both Presidents Johnson and Nixon strongly favored the idea, as did their secretaries of the interior. National recreation areas were initially proposed for New York and San Francisco—Gateway NRA in New York and Golden Gate NRA in San Francisco. President Nixon saw the two as demonstration projects for state, county, and municipal park programs. He called them Gateway East and Gateway West, hoping to create the impression of a single project. Other states soon demanded national recreation areas, however, and new NRAs followed in the Cuyahoga Valley south of Cleveland, on the Chattahoochee River in Atlanta, and in the Santa Monica Mountains of Los Angeles (Foresta 1984).

Purposes and Constituencies

The foregoing discussion makes clear that parks differ in character and purpose. While municipal parks mainly provide recreation, national parks enshrine places important to the national identity. Yellowstone and other natural parks preserve symbolic landscapes. Many patriotic and historical themes are encoded in the great western parks: discovery and exploration, conquest, the frontier and westward expansion, nature and wilderness values, national grandeur, rugged individualism, and so on.

Although they deal with built environments, national heritage parks like Independence National Historical Park and the Statue of Liberty and Ellis Island National Monument have similar missions of preserving national symbols and educating the public about historical events. These places are not really parks, certainly not in the Olmstedian sense: they do little to provide for leisure recreation in the tradition of a park. But almost from the beginning the National Park System has preserved historic sites and interpreted them for the public.

Gateway National Recreation Area represents a kind of hybrid of national and local park. NRAs preserve significant environmental resources, but they resemble municipal parks in emphasizing recreation. These parks bring the resources of the National Park System to urban populations who, it is thought, would not otherwise have national park experiences. Rather than reserving a contiguous space solely as a park, the NRA typically consists of noncontiguous collections of separate properties, including surplus military installations, nature reserves, and sites formerly operated by local park agencies. Gateway also includes residential "inholdings"—e.g., communities such as the Breezy Point Cooperative that proved too politically difficult for the Park Service to acquire.

New York City, facing extreme budget pressure in the early 1970s, was eager to transfer as many park facilities as it could to the National Park Service in forming Gateway. In addition to Riis Park, the mayoral administration of John Lindsay hoped to hand Coney Island over to the federal government. Riis Park had fallen into disrepair by the early seventies, and the city was glad to turn it over. Early plans drawn up by the Park Service prior to the inception of Gateway called for clearing away nearly all the structures and roads on Breezy Point, including Fort Tilden, the private Breezy Point Co-op, and the aging facilities of Riis Park. Planners envisioned a whole new recreational park on a large scale: an expanded swimming beach, an amphitheater, golf courses and playing fields, an environmental education building, parking lots, and "creative open space." New ferries would serve two ferry landings, with broad promenades leading to the beach (Foresta 1984).

The 1979 plan adopted several years later showed a much diminished vision. Local white communities in the Rockaways and in nearby areas of Queens and Brooklyn were resistant to plans that envisioned large new visitor populations. The sixties had been years of rapid neighborhood succession, and white homeowners in local neighborhoods, protective of turf and property values, opposed anything that might bring people of color in large numbers to Breezy Point. The Breezy Point Co-op proved to be too politically influential to be condemned, as envisioned in the earlier plan. Thus the vision at Gateway soon shifted from large-scale recreation to the quieter, less provocative virtues of historic preservation and environmental education. Rather than attracting unstructured recreational visits, the new plan called for supervised group experiences. The surplus structures and grounds of Fort Tilden would be preserved. Jacob Riis Park would continue on, its grounds and buildings also subject to historic preservation. The new emphasis on preservation and on quality experiences seemed to provide an alternative to the politically unreachable goal of mass recreation in an urban national park. It was strategic in being far more acceptable to the local communities, and it would require no increase in transit capabilities—which had proved similarly challenging (Foresta 1984).

The Politics of Funding and Service to Park Users

National parks operate in a political context very different from that of municipal and regional parks. Federal funding makes national parks generally much less dependent upon or responsive to local political conditions. Most municipal parks depend on local public funds; even a park like Prospect Park, whose privatized management raises substantial funds for the park, taps mainly city and state government sources. If parks are important to local constituencies, moneys for reconstruction and acceptable levels of operation will be found. The New York City Department of Parks and Recreation in the 1990s raised the capital funds necessary to reconstruct much of the recreational infrastructure at Orchard Beach.

The fate of Jacob Riis Park, Orchard Beach's cousin, offers a useful case study in how the politics of park funding can break down. In the 1960s and 1970s, these public beaches lost their original white middle-class constituencies. Like local parks in many cities at the time, they also lost funding, suffered from crime and vandalism, and ran down. Including Riis Park within the new Gateway National Recreation Area seemed to be its salvation: it would then be maintained with the resources of the National Park System and integrated within a coordinated system of recreation sites, historic landmarks, and natural areas, all available to public visitation.

As it turned out, federal resources, although forthcoming, were not adequate to reverse the deterioration of Riis Park. Despite the efforts of local managers, the park has no functioning bathhouse, some of the recreational facilities like handball and basketball courts are unusable, portions of the railing along the promenade are in a state of advanced decay, the grounds are worn and littered, the park lacks a planting program to replace dying trees, and attendance is down since the 1970s. Orchard Beach, by contrast, has recovered nicely from the municipal neglect of the sixties and seventies.

The contrast between Riis Park and Orchard Beach also illustrates how different management structures respond to changing demographics among users. On the one hand, Riis Park reflects the strengths of the National Park System in its planning processes. In ethnographic research alone, Gateway commissioned an ethnographic needs assessment for Riis and other park units in 1995, followed in 2000 by the rapid ethnographic assessment procedures (REAP) we conducted for Riis Park. NPS also has rigorous yet innovative criteria for protecting historic cultural resources. On the other hand, Riis Park has not adjusted well to the needs of some of its newer users, especially the large proportion of Latino visitors who use the grounds for picnic sites. These users find at Riis something like the tree-shaded margins of the tropical beaches in their home countries. Generally they would like to see more tree-shaded space made available at Riis Park for picnics. The NPS, however, operating under a historic preservation mandate to preserve the decorative intent of the 1930s planting plan, has restricted access to some of the landscaped grounds.[2]

It would appear that park hybrids such as Gateway may not work as well as hoped. Funding within the National Park System is allocated by priority: important parks with national constituencies rank highest. A park like Jacob Riis Park, which has no constituency outside New York City, ranks lower. There have been calls to return Jacob Riis Park to the city park system, where it would have relatively greater stature. The newer Boston Harbor Islands National Park Area follows a different, perhaps more workable model. There, instead of having the National Park Service take over territory formerly owned by city and state agencies and private conservation trusts, those parties remain owners and managers with the Park Service providing a management overlay of research, planning, coordination, and public education and outreach.

Municipal parks, for the most part, show more flexibility in responding to changing user needs. Management at Pelham Bay Park, for example, sees the cultural expressions of Latino park users as part of the park's identity, and thus as worthy of encouragement and support. Pelham operates within a local political context, a city and particularly a borough (the Bronx) with a large and influential Hispanic electorate. In a sense, the political system delivers certain public goods for this group of citizens, including a rehabilitated seaside park. Although the local parks department in New York City lacks the ethnographic research program of the National Park Service, park management can be pragmatic and flexible in making Orchard Beach responsive to the cultural associations of its predominantly Latino constituency.

Conclusion

This brief survey of parks demonstrates the variegated nature of urban parks. Recreation is always an important park value, but the presence of other values makes the landscape of park purposes and uses considerably richer. Even local parks exhibit important differences in use and character, and in their histories, that make for complex management issues. For example, the history of "laissez-faire" management at Pelham Bay Park is reflected in the present management's willingness to allow user groups to at least temporarily appropriate spaces and facilities for unofficial uses that nevertheless contribute to the park's ability to appeal to competing constituencies. Prospect Park's history of strong design control underlies the present management's efforts to emphasize the park's design integrity and to restrict uses inconsistent with the design philosophy of passive, restorative, and democratic recreational activity.

Urban national parks, like all national parks, operate under explicit, publicly disseminated management objectives. These generally balance the competing goals of recreation and preservation, usually making recreation values secondary to the various preservation interests in any given park. Among national heritage sites like Independence and Ellis Island, preservation and interpretation of historic values is surely the management priority. Yet Independence is used by some local residents and downtown office workers for passive recreational activity, such as walking and sitting, eating, meeting friends, and so on. The national recreation areas like Gateway were established to open up scenic waterfront areas to metropolitan recreation, yet they increasingly emphasize preservation of relict natural environments and interpretation of historic structures. Even at Gateway's Jacob Riis Park unit, historic preservation issues and local politics complicate the park's recreational mission.

One theme to keep in mind while reading the case studies is the contest between the official and the formal, on the one hand, and the vernacular, on the other. Vernacular uses are rarely noticed in official park descriptions, but they play an important role in the social and cultural life of parks and in management issues. They often have an unexpected vitality that energizes formally designed parks. At Jacob Riis Park, picnicking under the shade of trees revitalizes the park but also causes consternation on the part of managers seeking to preserve the historic scenic values of the park's landscapes. In Prospect Park, management has often turned to the historic design as a vision for park regeneration, always struggling against the contemporary vernacular of soccer and volleyball games, mountain biking, and other activities inconsistent with the historicist ideal. Many parks, and all the ones studied in this volume, have strong formal qualities that resound with the vitality of contemporary urban uses. What we hope to do in this book is examine how the formal and the vernacular come together to produce the unexpectedly vital but fragile cultural alchemy of contemporary urban parks.

Notes

1. Bushnell Park in Hartford was the first American city park to use the principles of English landscape design but Central Park was first in its class of parks of at least 100 acres.

2. The NPS has acknowledged the demand for picnicking space by converting a disused ball field to a picnic and cookout ground and building several pavilions to provide shade as well as shelter from rain.

Chapter 3
Prospect Park
Diversity at Risk

Introduction

In their sociability and informal layout, places of working-class recreation continue to resemble the vernacular weekend resort, or "grove," that lay outside every nineteenth-century American town. This was an open space with trees, fields, and water at hand, used informally for recreational gatherings by the townspeople on Sunday afternoons (Jackson 1984). Although such places have yielded to urbanization and to the evolution of leisure time activity, parts of Brooklyn's Prospect Park seem much like the old grove. For instance, on the Peninsula lies a pleasant field of two or three acres bordering the lake on one side and a placid stream on the other. A dirt path meanders along the shore toward the woods beyond the field. Families gather for picnics under the trees or to sit and look out at the water. Men and boys fish, young people play ball, and children ride their bikes. Groups of adults and teenagers stroll along the nearby road, pausing to look out at the water. This is a favorite place to bring dogs because they can run in the field and swim in the lake. The current Prospect Park Administrator brings her dogs here early on a weekend morning, as she once told us.

This place has become a setting for the sort of vernacular recreation that typically either escapes official notice or is taken for granted. Few of its users spend their time here thinking about Nature or Beauty (although it is popular with bird watchers), but the pleasant natural setting is essential to the leisurely recreation that goes on here. It is spacious, easy on the eye, and soft underfoot. Birds and ducks occupy the lake surface and the air above, and fish live in the water. Big trees cast shade, and when one occasionally falls over its trunks and branches make an attractive jungle gym. This place has so far escaped the restorer's attention: nothing is torn up, fenced off, dug up, or reconstructed.[1] The scene is not fussy or pretty in any artificial way; it lacks gardenesque or horticultural effects or picturesque structures. There once were such structures and visual effects at this location, but natural processes have had their way as tastes changed and maintenance budgets declined. Now the field seems just to exist, like a quiet rural place unclaimed by urban development.

Urban landscape parks may be civic ornaments and nature preserves, but they are also vital social spaces. Places like the Peninsula field provide settings for an informal coming together of disparate groups in public spaces where the natural environment engages people's attention. The purpose of this chapter is to contrast the park as a social and cultural space with its socially constructed identity as a historic landscape and "last forest in Brooklyn" in need of protection and restoration. Landscape parks like Prospect Park are indeed historic and worthy of protection, yet there is too little understanding of how the disruptions brought about by restoration can affect parks as social spaces. We need to ask whether the management's course serves users well, especially as Prospect Park is at the forefront of a national movement toward private funding and management of public parks. Private groups are responsible only to their memberships and boards of directors, not to the public at large.

Much of the existing ethnographic research on parks has been at the behest of park managers who need specific kinds of information about users rather than general knowledge. Such is the case of the Prospect Park User Study of 1996–1998, the third such study commissioned by the park management since 1980. Yet the user study data provide a basis for a much broader appraisal of user values: not just activity and behavior, likes and dislikes, but also the character of people's park experience and the meanings the park has for them.

Methodology

The Public Space Research Group conducted the Prospect Park User Study beginning in summer 1996. Data collection continued over the following year; the final report was submitted in May 1998. Our proposal to the park administrator emphasized the potential for the user study to explore how the park functioned as a sociocultural space. The proposal called for close-up ethnographic work in selected areas of the park, and about half the final report was devoted to these mini-ethnographies.

We explored user attitudes in depth through participant observation and 357 user interviews. We made a great effort to find differences in park values among the user population, basing our analysis on variation in the data across demographic categories such as race/ethnicity, age, sex, income, and level of education.

In an era of dwindling public resources, park officials had decided that changing user attitudes and behavior could substitute for the traditional level of maintenance and supervision that park budgets no longer allowed: the users would become "stewards" of the park. Management wanted to educate users about the park's artistic legacy and its biological environment. Park officials also wanted to attract users to the interior areas of the park to ease overuse of the perimeter. As the Prospect Park Alliance wrote in its 1995 grant application to the Lila Wallace Reader's Digest Fund,

> [P]ublic behavior must be transformed.... It never occurs to most Park visitors that riding a dirt bike up the hills of the Ravine will harm tree roots.... It simply never occurs to most Park visitors that placing a charcoal grill directly under a tree will surely contribute to its death.... The challenge is to learn more about the Park's users, their needs, and what they do and want to do in the Park. From that information, programs can be designed that will help them use the Park in a more constructive way.

To this end, park officials were very interested in finding out where users come from, how often they come, how long they stay, and how long they have been coming. They were interested in some aspects of people's knowledge of the park and their attitudes toward it, such as the places and programs frequented, fears, and complaints about facilities and services.

One of the uses of a user study lies in identifying constituencies that can then participate with management in planning and programming. The management of Prospect Park already has extensive contact with user groups. Among the most vocal and socially well connected are the dog walkers. As a group, they have organized politically to negotiate a favorable dog policy in return for a substantial measure of self-policing and managing of dog wastes. The park is also in contact with groups such as soccer clubs to modify their use of the park. They negotiate rotating of play areas with the clubs so as to reduce the impact of soccer games on any given lawn area. Management has also cooperated with some user constituencies in participatory planning: an example is its work with an informal group of drummers (about which more later), first on a temporary relocation pending reconstruction, and then on a new layout of their traditional location.

The user study included an extensive quantitative presentation and analysis of the interview data. We coded and counted all the answers to the interview questions. All the responses were presented in frequency distributions. We then cross-tabulated many variables to find statistical variation among the participant sample.

The user study included a census of visitors. Pedestrians entering the park at all 18 entrances were manually counted at particular times agreed on in advance. The counts were then statistically extrapolated over the whole year, which produced an annual usership estimate of approximately 5 million. Based on projections from a user study completed in 1987, Prospect Park was already claiming 6 million annual visitors and has continued to use that figure in press releases and other public information efforts.

Park managers asked the census takers to record whether the persons they counted were white, black, or Hispanic, based on appearance. Although this methodology has obvious flaws, in that people's identities are not always what they seem, the resulting breakdown was consistent with earlier estimates and with our own ethnographic observations: roughly one-third white, one-third black, and one-third Hispanic.

Interview Sample

The interview sample of 357 park users had 119 white participants, 102 Hispanics, 117 blacks, and 13 Asians. The sample was a stratified opportunity sample that sought a balance of blacks, whites, and Hispanics in keeping with existing estimates of the racial profile of the user population. Of the participants, 153 were of either low or lower-middle income, and 123 were in the middle and higher-income ranges. Almost two-thirds of the interview sample had walked to the park; the rest drove or took public transportation. White participants were more likely to live within easy walking distance. Most of the participants who had traveled for more than a few minutes were people of color. The best-represented neighborhoods in the sample were Park Slope (24%), Flatbush (13%), Kensington (10%), Crown Heights (8%), and Prospect Heights (6%), which are all adjacent to the park.

Research on Race and Ethnicity in Parks

Researchers working toward social justice are almost naturally drawn to investigating racial and class differences in multicultural urban settings. These efforts typically find resources to be unevenly distributed, though not necessarily as a result of deliberate park policies. Much research on parks, including the case studies in this book, is funded under contracts with park authorities. Managers as research clients sometimes find it safer to avoid addressing issues of race, ethnicity, or social class directly. In such cases the research instead takes a less controversial approach—for example, sorting users by location or neighborhood. Jeff Hayward (1990) conducted a telephone-based survey of residents of neighborhoods bordering on Franklin Park in Boston. The survey sought to find out how neighbors felt about the park, whether they used it, and their thoughts on how to improve the park. Hayward's report analyzed the results by neighborhood without addressing the demographic characteristics of the respective neighborhoods. Hayward did a similar study later for Chicago's Lincoln Park.

In more homogeneous societies, race and ethnicity may not be an appropriate analytical framework. In her research in San José, Costa Rica, Setha Low (2000) sorted the park user population by gender, class, and whether they were Costa Ricans or North American tourists. These categories made sense in that setting and in terms of Low's ethnohistorical analysis of the meaning of the plaza in Latin American cities. Park user populations can also be defined as constituencies aligned according to type of activity. In Chapter 8 of this volume, Setha Low discusses her use of a constituency-based analytic framework for social research in parks.

A body of research in the leisure and recreation field explores the apparent differences with respect to park choices and values among users of different social class, race, and ethnicity in the United States. While much of the work involves wild lands rather than urban parks, reviewing it is useful for understanding the treatment of equity issues across race, ethnic, and class boundaries. Park "values" may be defined as "the symbolic content attached by a group" to objects or place (Washburne 1978, 177). Relative to the national population, national park users have been found to be disproportionately white and middle class (Washburne 1978; Woolf 1996; Taylor 2000). "Wild land resources seem to be largely the domain of white America" (Washburne 1978, 176). African Americans, Asians, Hispanics, and other people of color have been found to attend only local parks in numbers proportional to their percentage of the total population (Woolf 1996).

The literature discusses two alternative theories to explain the differences in participation between whites and people of color: marginality and ethnicity (or subculture) (Washburne 1978; Hutchison 1987; Floyd et al. 1994). Marginality explanations for underparticipation of people of color in wild land resources focus on poverty and socioeconomic discrimination. They reason that more people of color lack cars and cannot otherwise afford the cost of visiting national or nonurban state parks, suffer disproportionately from unmet basic needs, work longer hours than whites, and so on. The marginality view implies a policy aimed at increasing minority access to wild land parks. Urban national parks like Golden Gate in California and Gateway in New York and New Jersey—parks that seek to bring a national park experience to city dwellers—were established in response to marginality concerns (Woolf 1996).

Advocating an "ethnicity" alternative, Randel Washburne (1978) argued that blacks have different cultural values from whites in relation to wild lands. Blacks, he wrote, maintain their ethnic status by socializing with one another locally, in their neighborhoods, churches, and other community institutions, as well as in local parks. Washburne cited California survey data showing that, relative to whites, blacks prefer playing basketball, attending spectator sports events, and community and neighborhood activities; are equally inclined toward fishing, hunting, and crabbing; and are much less inclined than whites toward "trips and vacations," "walking, hiking, climbing," and "visiting regional or remote parks." For Washburne, the ethnicity explanation for black underparticipation in wild land parks supports a shift in policy away from a standard wilderness-based vocabulary to providing a variety of public spaces, including local recreational facilities.

Some park management officials agree with Washburne. For example, Roger Kennedy, director of the National Park Service during the Clinton administration, stated that immigrants from Africa, southern Europe, Southeast Asia, and Latin America have strong traditions of family and clan gatherings in village squares, city parks, and orchards close to home (Woolf 1996). Kennedy goes on to ascribe whites' attendance of western national parks to a North Sea tradition of traveling to distant natural areas for vacations of camping, hunting, and fishing. Perhaps so, but people of color also have traditions of long-distance travel. Ethnicity alone cannot entirely account for their low attendance of wild land parks.

A more compelling explanation stems from the likely impact of racial hostility on park use participation by people of color. West (1989) found that black residents of Detroit were more likely to use city parks than white city residents, who visited suburban parks proportionately more than blacks and visited them more often. West argues that the difference is not explained by cultural group (ethnicity), because whites and blacks expressed equal levels of interest in using metropolitan parks. Marginality did not explain the difference either. Instead, West attributes black underrepresentation in suburban parks in large part to their perception of a potentially hostile social environment. Sometimes the discrimination is overt, as with suburban Dearborn's effort to prohibit nonresident use of its parks (at a time when Dearborn, Michigan, was predominantly white). Even without overt discrimination, a black family may think twice about visiting a mostly white regional park located in the white suburbs.

Racism figures similarly in William Kornblum's (1975) assessment of people-of-color underrepresentation in western national parks. On the basis of studies he conducted for the National Park Service, Kornblum notes, as important factors, the whiteness of both park staff and other visitors; the prospect of a long drive through rural, white regions of the country to get there; and the cost of the trip (Woolf 1996).

Taylor (2000) observes that people of color do visit wild lands and some have reported unpleasant encounters with white visitors, where they were "stared at, stared down, and stared out" of these areas. "White users and wild land managers," Taylor argues, "assume that park, forest, and wilderness users will be white and that the wild land areas are exclusive white spaces" (Taylor

2000, 174). Taylor suggests using more images of people of color in parks to help diversify these spaces. She advocates pulling away from the typically white, urban, middle-class view of wilderness as "empty, virgin land, untouched by human hands, where [whites] can retreat from urban problems and people" (Taylor 2000, 174).

Researchers find racial and cultural patterns in local park usage as well, as to both the choice of park and preferred activities. A study of parks in New Haven, Connecticut, found that black residents were in general attracted to more social facilities, like ball fields and picnic grounds, and whites to tennis courts and jogging trails (Taylor 1993). Blacks preferred one city park disproportionately because it had safe play equipment and fields to play on and because there were other blacks present. Taylor argues that ethnicity explanations rely on a too-narrowly-specified model of race, that is, "blacks," "Hispanics," and "whites." In the New Haven study, for example, African Americans were attracted to one park for its overall peaceful atmosphere, whereas Jamaicans were more likely to be attracted by certain facilities in the park.

A study of Chicago's Lincoln Park (Gobster and Delgado 1993) found that whites visited the park predominantly alone or in couples, while black and Hispanic users came in larger groups of family and friends. Black and Hispanic users also engaged in passive social recreational activity more than whites—activity like picnicking and talking and watching organized sports. In another study of Lincoln Park, Hutchison (1987) reports clear differences between Hispanics and both blacks and whites. Hispanics shared with blacks and whites a preference for walking and bicycling but were much more prone than the others to playground use, picnicking, watching sports events, and lounging on the grass. In our work on large municipal parks in New York City, the authors' interest in making cultural knowledge a basis for park policy thus led to an analysis of the data that included differences in race and class.

Historical and Social Context

Prospect Park occupies a site of 526 acres in north-central Brooklyn. It was built between 1867 and 1873 after many years of advocacy from Protestant ministers and other civic leaders in the then fast-growing city of Brooklyn. Prospect Park is part of a civic complex that includes the Brooklyn Botanic Garden, the Brooklyn Museum, and the main building of the Brooklyn Public Library.

Prospect Park and its connecting parkways—Eastern Parkway and Ocean Parkway—were planned to spur high-grade residential development in the then open lands nearby. That strategy was very successful as Prospect Heights, Crown Heights, Park Slope, and Flatbush all grew up into densely packed but

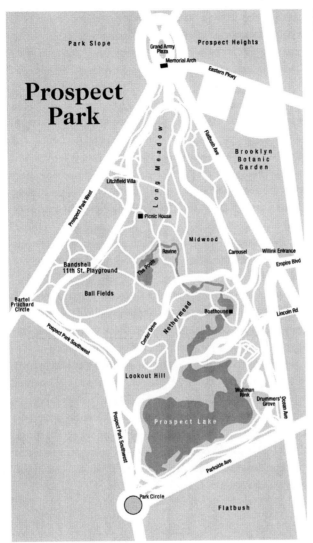

Map 3.1. Prospect Park

prosperous residential sections. These and other neighborhoods around the park have remained heavily residential: Brooklyn's industrial and office districts are all some distance away. The adjacent neighborhoods each have an immediate spatial relationship with their side of the park, and the park's ethnic makeup and general atmosphere changes from one neighborhood's zone of influence to another.

Landscape Design

The park was designed in the pastoral landscape style, which emphasizes calming scenery of meadows and pastures that rise over gentle hillsides, spreading trees, woods in the distance, and ponds and streams nestled in the valleys. Its acreage includes a rolling upland pasture area, a region of wooded hills, and in the southern third, a flat area on lower ground that the park's designers dug out to make a shallow 60-acre lake. Prospect Park's major pastoral composition is the Long Meadow, a 90-acre area that occupies much of the west side of the park. Pastoral scenery is also found in the Nethermead, a smaller open area more in the center of the park, and in the clearings and groves around the lake shore. The park's fields and meadows were supposed to have the coarse turf of real pastures, and indeed sheep grazed on them in the early years. Today they are widely used for picnics and cookouts, ball games of various kinds, kite flying, sunbathing, dog walking, and running.

Olmsted and Vaux sought to keep elaborate flower beds and other artificial planting effects out of their pastoral composition. Still, the park was a horticultural showplace in the natural style that featured flowering shrubs and ornamental trees planted along the paths and drives, at the entrances, and at the numerous bridges and overpasses. Few plantings of this type survive today; much of the open ground resembles the vernacular North American park of grass and trees—a form based on the pastoral style practiced by Olmsted and Vaux but simplified so that little maintenance beyond routine mowing is required (Jackson 1984; Wilson 1992).

As in many local parks, maintenance at Prospect Park has been uneven. Years of neglect by the city have left eroded hillsides, overgrown and littered woods, broken pavements, and water bodies clogged with weeds and silt. In the Olmsted and Vaux design, the woods were open groves of trees, shrubs, and picturesque water features, furnished with gazebos and hospitality facilities to attract picnickers. Now the woods have thickened with the overgrowth of aggressive exotic species, and more recently of indigenous colonizers like black cherry. The balance intended of conifers and deciduous trees, large and ornamental trees, and trees and shrubs has changed to denser stands of visually undifferentiated hardwoods. Thick stands of trees have spread into many areas intended to be visually open, blocking views and limiting the space available for recreational use.

As designed, this environment was expected to have a civilizing influence on its users, who would behave much as the patrons of an expensive rural resort do today. Guests of the Mohonk Mountain House in New York State take leisurely

Figure 3.1. The Long Meadow in Prospect Park

rambles over the scenic wooded grounds, through the flower gardens, or along the lakeside paths. They can pass a pleasant half hour in one of the numerous rustic shelters overlooking picturesque Mohonk Lake, take a rowboat out on the water, or play croquet on the lawn. Prospect Park was planned to be very like Mohonk, only free of charge: a pastoral retreat with gentle meadows and wooded groves, picturesque waters, charming carriage drives, a hilltop overlook, and comfortable facilities for visitors designed tastefully to blend with the landscape. In this genteel environment working-class and immigrant visitors were expected to learn the social skills they were thought to need to better themselves and become good citizens. People would learn to behave well and to interact with one another; the park's genteel constituency would provide models of good sportsmanship and of how voluntary groups come together and interact in the public realm. Olmsted saw his parks as training grounds for citizenship.

Built Features

Prospect Park has several crowd-attracting features. The Bandshell, along Prospect Park West near the Ninth Street entrance, is the site of a popular, summer-long program of outdoor pop music concerts. On the opposite side of the park, along Flatbush Avenue near the Willink entrance are the Prospect Park Zoo, the Carousel, and the Lefferts Homestead, a historic house museum geared to children. There are six children's playgrounds at different locations along the edges of the park, each relating to the adjacent neighborhood. Farther inside the park, a cluster of baseball fields occupies the southern third of the Long Meadow, and an artificial skating rink borders the east side of the lake. There are several more ball fields in the Parade Ground, a separate area across Parkside Avenue from the park's southern border. Brooklyn's large West Indian community makes use of an area along the circuit drive near the corner of Ocean and Parkside avenues for a weekly drumming and dancing festival that can attract hundreds of participants, onlookers, and food and craft vendors on a warm Sunday afternoon.

Most of the existing "places of congregation," as Olmsted called them, are not original. The Bandshell and the zoo were constructed in the 1930s, and the Long Meadow ball fields and the skating rink were added circa 1960. None of the playgrounds were extant prior to 1940. Olmsted and Vaux's plans provided places for people to gather too, such as the Concert Grove, the Lookout, and the Refectory. The Refectory was never built; others (including the Concert Grove and the Lookout) were altered or undermined by later management decisions so that they never fulfilled their intended purposes.[2] Still others—for example, the "Dairy Cottage"—have disappeared.

Management

The park's current management describes itself as a partnership between city government and private advocacy. A group of park advocates and civic leaders, concerned about the park's decline in the 1970s and early 1980s, organized a management and advocacy entity known as the Prospect Park Alliance, in 1987. The Alliance "brings together the community, corporate, and government resources" necessary to maintain and renew the park, raising funds for landscape restoration and community programming and operating a volunteer program (Prospect Park Alliance Annual Report 1995). The park management consists partly of the New York City Department of Parks and Recreation and partly of positions funded by the Prospect Park Alliance. The Parks Department staff provides routine operations such as lawn mowing, trash pickup, and enforcement patrols. The Alliance appears to be the entity responsible for efforts to renew the park through fund-raising for landscape restoration, outreach and advocacy, and building user constituencies. The president of the Alliance, Tupper Thomas, also holds the city government position of Prospect Park Administrator. Ms. Thomas insists that the Alliance's role is only to support the city's management of the park and that all management decisions are made by the city, not the Alliance.[3]

The mission of the Prospect Park Alliance is park renewal in keeping with the park's "Olmstedian" origins. The Alliance has shifted the management emphasis from recreational facilities and decorative effects to restoring the health of the park's natural ecology, especially its woodlands. Thus the park management is encouraging the spread of woodland into some areas that were formerly open, to create more wildlife habitat, and planting young specimen trees at the margins of the Long Meadow and elsewhere. The intent is to restore some of the landscape complexity lost to the death of older trees and to the requirements of mechanical lawn-mowing methods. The Alliance is well into its centerpiece initiative, the 25-year Woodlands Campaign to restore the water system and adjacent woodlands in a project that mixes historic preservation and replication of the original landscape architecture with replenishment of soils and vegetation. Substantial parts of the interior woodlands have been fenced off from public access to allow this work to progress with minimal interference.

Since Olmsted's ideas and designs were laden with class-bound prescriptions for moral behavior, an effort to restore the park to its Olmstedian glory may impinge on some of the park's present-day uses and meanings. This chapter reveals some of the differences in values, both intergroup differences among users and between users and the policy of Olmstedian restoration.

Findings of the Study

The social life of Prospect Park is both rich and diverse. The park is a site of cultural self-expression for certain groups, including African/West Indian drumming and dancing, Haitian roots music, and an officially sponsored yet very diverse pop music concert program. Still, the most common activities are walking, sitting, and exercising. Dog walking is prominent, and the activist dog-walking constituency has achieved a relatively generous off-leash policy at certain hours and park locations. Many people use the park for picnics and cookouts. There are a variety of athletic activities—some officially sanctioned, namely softball and baseball on dedicated playing fields. Most other sports, including soccer, volleyball, and ultimate Frisbee, are tolerated but unofficial.

Much of the action in the park is common to people of different cultural and class backgrounds. Characteristic activities like walking, exercising, watching the ducks, and visiting the playground are well distributed among the different constituencies. There are also important differences among user groups, however. The emphasis in this chapter is on the differences in the data from one group to another—differences in values and differences in activity.

Figure 3.2. Sunbathers at Prospect Park

Figure 3.3. Winter day, Prospect Park

Values

The User Study interviews included the question "Does the park have any special meaning for you?" We coded the wide range of answers according to response categories based on the actual responses—categories like "release-refreshment-escape," "freedom," and "alternative to being in the apartment." The categories are presented in a frequency distribution, in descending order, in Table 3.1.

"Relaxing/tranquil" had the highest number of responses—58. "Appreciating nature/wildlife" ranked second with 56 responses. These two values are consistent with one of the generative ideas of the park—that people should find relaxation and tranquility in a spacious and scenic landscape setting.

"Appreciating nature/wildlife" reflects different ways of relating to the natural environment of the park, certainly more than just the choices of bird-watchers or other users of a naturalist bent. Many participants cited nature or the animal life of the park as something important to them—some with precision, others just with an emphatic "Yeah, it's nature!" Surely, enjoying nature in some sense is part of the conscious experience of the many participants from the above discussion who gave walking in the park as an activity. "Appreciating nature/wildlife" reflects the comments of bird-watchers as well as those who feed the ducks.

Other highly ranked values included childhood memories and family memories, "release-refreshment-escape," "place to recreate," and "nice" and "beauty." The category "release-refreshment-escape" reflected comments such as getting away from the city or finding release from stress. "Relaxing/tranquil," the highest-ranked category, is similar in meaning to "release-refreshment-escape" but lacks the implied comparison to city life. "Place to recreate" was the code reserved for comments about enjoying the park for exercise routines and team sports.

Childhood and family memories associated with the park were always favorable and seemed fundamental to the experience of many users who had come to the park as children. "Place memory," in which the park reminds the person of some other place, was important to a number of immigrants and other participants who had grown up somewhere else. "Nice" and "beautiful" are related values of judgment, "beautiful" referring more explicitly to aesthetics, and "nice" a term of more general sensory pleasure. These comments were often accompanied by some reference to the spaciousness of the park, coded as "large open space." "Place to spend time," "alternative to being in the apartment," and "habitual use/second home" all reflect values that seemed to be about using the park as a second home or outdoor living room. "Community-

Table 3.1. Park-Related User Values in Prospect Park

Category	Count	Percent of Responses
Relaxing/tranquil	58	9.1
Appreciating nature/wildlife	56	8.8
Childhood memories	36	5.6
Other	36	5.6
Release-refreshment-escape	35	5.5
Nice	32	5.0
Place to recreate	28	4.4
Family memories in general	28	4.4
Beauty	27	4.2
Place memory	25	3.9
Place to meet friends	21	3.3
Large open space	21	3.3
Place to spend time	20	3.1
Community/public resource	20	3.1
Alternative to being in the apartment	18	2.8
Nothing/none	18	2.8
Inspiration	15	2.3
Romantic memories	15	2.3
Ethno-cultural association and identity	12	1.9
Habitual use/second home	12	1.9
Memories of activities w/ friends	12	1.9
Fun	11	1.7
Place to visit on the job	11	1.7
Place to feel safe	11	1.7
Parenting memories	7	1.1
Place to be alone	7	1.1
Freedom	6	0.9
Good fishing	5	0.8
People are friendly	5	0.8
Only place things can happen	4	0.6
School-related memories	4	0.6
Social diversity	4	0.6
Hanging out	4	0.6
Place to be when not working	3	0.5
Did not ask	3	0.5
Refused to answer	2	0.3
No answer	9	1.4
Total responses	*640*	*100.0*

Table 3.2. Park Values in Prospect Park User Study Reclassified

Category	Count	Percent of Responses
Place to go/things to do	129	21.9
Relax/release	93	15.8
Nice/beautiful	74	12.6
Nature/wildlife	56	9.5
Memories	51	8.7
People	42	7.1
Other	36	6.1
Safe	32	5.4
Place memory	25	4.2
Civic resource	20	3.4
Nothing	18	3.1
No answer	9	1.5
Did not ask	2	0.3
Refused to answer	2	0.3
Total responses	*589*	*100.0*

public resource" was the code for comments to the effect that the park contributes importantly to the sense of Brooklyn or New York City as a community of citizens.

To find demographic variation in the data we collapsed the individual response categories in Table 3.1 into a set of broader groupings. Table 3.2 shows the new grouping. The new classification is helpful in seeing demographic patterns in the data, presented below in Table 3.3. The cross-tabulation demonstrates that people of different ethnicity share important park values. "Relax/release" is quite evenly distributed across census groups, as is "nature/wildlife" and "memories."

There are also interesting differences in the data. The category "places to go/things to do" refers to values (from the original variable) such as "place to recreate," "large open space," "place to spend time," "alternative to being in the apartment," "good fishing," and so on. The meaning of the new category is of the park as a place of opportunity to do things or just a place to be in. It has a higher incidence among blacks and Hispanics than among whites. The value "people," which includes "ethno-cultural association and identity," "place to meet friends," and "social diversity," has a higher incidence among blacks. Conversely, the category "nice/beautiful" is associated with whites. "Civic resource" is a "white" and "black" value. Whites were less likely than others to call the park "safe."

Table 3.3. Values and Census Group in Prospect Park User Study

Value	Responses by Census Group			Total
	White	Latino	Black	
Place to go/things to do	28	43	54	125
Relax/release	29	26	32	87
Nice/beautiful	37	10	21	68
Nature/wildlife	20	14	17	51
Memories	15	17	19	51
People	11	9	21	41
Other	18	11	7	36
Place memory	13	4	8	25
Safe	5	14	12	31
Civic resource	10	0	6	16
Nothing	28	8	3	39
No answer	29	2	3	34
Did not ask	37	1	0	38
Refused to answer	20	2	0	22
Total	*300*	*161*	*203*	*664*

User Activity

Participants were asked what activities they were pursuing that day. "Walk in the park" was cited by one-third of the interview sample (122 participants). Walking is one of the central activities of a landscape park and a perfect fit with design intention: parks like Prospect Park were designed with walking in mind. Other high-ranking activities were "visit playground," "relax," "picnic/barbecue," and "to be with family/kids."

Although the data show the wide range of activity to be fairly evenly distributed across education level and ethnicity, there were—as with values—some interesting differences. For example, few whites were picnicking and barbecuing: Of 49 participants engaged in picnicking or barbecuing on the day of the interview, 27 were Hispanic, 16 were black, and 6 were white. Perhaps more of the white users have their own yards or second homes or easier access to other open space through higher automobile ownership (none of these were questions we asked). Hispanic users were more likely to see their park experience as something social rather than a matter of private relaxation. Of the 63 people who said they were there to relax, only 8 were Hispanic, in contrast to 19 white and 36 black. Of 14 participants who said they were hanging out, 10 were black, 2 white, and 2 Hispanic. Whites were more likely than others to say they were walking: among the 121 participants who cited walking as an activity, 51 were white, 37 black, and 24 Hispanic. In addition, most dog walkers were white. There were very few Hispanics among the runners and in-line skaters, although of 30 cyclists interviewed, 7 were Hispanic, 9 white, and 14 black.

The following paragraphs provide examples of the uses and meanings of the park characteristic of these broad ethnic categories: black, white, and Hispanic. These ethnographic descriptions—of cultural festivals, picnicking, and enjoying nature—serve also as examples of how Prospect Park succeeds in accommodating the various activities of its diverse constituencies.[4]

CULTURAL FESTIVALS

It is on the east side of the park that the contemporary Brooklyn of immigrant groups and diverse peoples of color makes its presence felt the most. To paraphrase one visitor, West Indians run the show here. The southeastern corner of the park, near the Parkside-Ocean entrance, is a focal point for West Indian and African American cultural activity. A local folk artist carved human images into a tree stump by the lakeshore; this site, the "Gran Bwa" (a corruption of *"grand bois"*) or "head," became a place where men gather to play Haitian roots music. The stump has decayed since the research and the carved image is no longer recognizable. It was an icon for Haitian people (the name, from Haitian mystical tradition, belongs to the spirit of the woodlands), and Haitians still gather there on the logs that circle the old stump. Haitians also gather in the Oriental Pavilion every Sunday and give informal concerts to appreciative Haitian audiences.

Surely the most prominent "grass-roots" cultural attraction in the area is the drumming phenomenon that takes place on Sundays in an otherwise nondescript place along the East Drive. Following is an excerpt from field notes taken by Charles Price on July 14, 1996.[5]

> The sound of the drums is audible from some distance. Approaching the site, one sees vendors selling foods, drinks, and arts and crafts goods. The actual drummers' site is simply a group of benches arranged in a U-shape around a patch of ground worn flat and bare by the heavy traffic of feet.
>
> At approximately 3:00 p.m., there are from 75 to 100 people gathered right around and on the benches; in the peripheral area outside the benches another 200 or so people mill around. By 5:00 there are twice as many. In this peripheral area people are watching and listening to the drumming, drinking and eating, smoking, talking and dancing. People mill around the many vendor tables looking at and buying merchandise and many buying food. There is no place right nearby to sit so many who buy the chicken, corn on the cob, and other offerings eat and

drink it standing up. There are several drummers and percussionists in this peripheral area, some of them playing their instruments and others resting.

Some passersby, many of them jogging, in-line skating, and cycling along the circuit drive, pause along the roadside to observe the scene. Some people have set up blankets and tents in the area, and are having what might be called picnics. There is a noticeable presence of foot and cruiser police in the area, as well as at least two park enforcement patrol officers, both black women, armed with nightsticks. The crowd gathered at the site is quite ethnically and nationally diverse. The markers and clues that suggest the participants come from places such as the Caribbean (Guyanan, Jamaican, and Trinidadian flags are on display); Africa (people wearing West African garb are evident, as are the sounds of French and Wolof); Latin/Central America/Spanish Caribbean (Spanish being spoken); African Americans, and some whites.

The Drummers In Action. The drums in use range from large and small bass drums, to congas, bongos, and talking drums. Other percussive instruments abound, such as shakerees, tambourines, bells, and a few people have recorders and flutes. The drummers are also a multiethnic mix of people, most of them black men. Most of the dances going on in the circle are Yoruba movements. The whole endeavor is quite participatory; all you need is an instrument of some kind, or an urge to dance. No one person guides or directs the drumming. Generally, one or another drummer starts what may be called a "baseline" rhythm, often one of the bass drummers. As this central or baseline rhythm becomes consistent, other players pick up the rhythm and perform whatever variations on the central rhythm they wish. Thus the cacophony soon becomes a syncopated rhythm. As more and more people "catch" the emergent rhythm, the sound and intensity of the drumming increases until a "groove" is reached. They then hold this "groove," especially if the onlookers and dancers are intensely involved. Eventually, the rhythm decreases in intensity until it stops entirely, or until only a few drummers are left playing. If these few drummers continue playing, then they often set the tempo for a new rhythm, which begins in the same way—other drummers slowly begin to fall in line with the new baseline or central rhythm.

Most of the drumming rhythms are of an African variety, mainly Nigerian, but some rhythms sound Haitian, and some seemingly of an Afro-Brazilian variety. On this particular day there are no Rastafarian drum rhythms, nor any other Jamaican rhythms. The reason may be

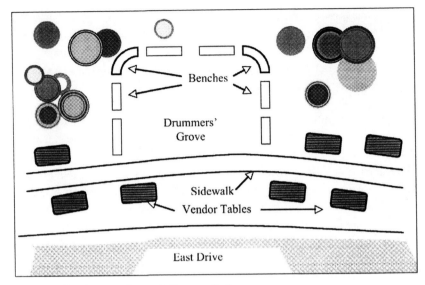

Figure 3.4. The drummers' grove in Prospect Park

that most Rastafarians avoid direct participation in activities that they, or Jamaicans in general, call "Obeah" or "science." These are folk terms for what others might call witchcraft or Vodun. The drum beats audible today are often associated with such practices, and the offerings in evidence corroborate this conclusion. All in all, as it seems today, the drumming is not simply a gathering of musicians, but also an event of religious and cultural content—even of religious and cultural significance.

Most of the people interviewed here come to the drummers' grove to hang out as well as to listen to the drums. Here they can eat, socialize, watch people, and examine the arts and crafts goods for sale. Several people said that they come to support the drummers. Some Sundays, depending on weather, the drumming sessions can go until 10:00 or later, and people remain in the area well past midnight.

Many of the African American and West Indian participants in the user study knew of the drumming event and spoke of it favorably as a cultural tie to the park. Not everyone in park management was as appreciative, some remarking that the intensive weekly activity would weaken the trees in the area from soil compaction. However, when the grounds around the Parkside-Ocean entrance were reconstructed in 1999 and 2000, management worked with the drummers to continue the drumming tradition. They first agreed on a temporary alternative location and then worked with the drummers to plan their return to the grove once reconstruction of the surrounding grounds was finished.

PROSPECT PARK

PICNICKING

Picnicking and cooking out occurs in many locations in the park. Irrespective of location, the picnickers are overwhelmingly people of color, usually black or Hispanic. There are picnic grounds at Parkside-Ocean, at Ninth-Eleventh streets, and in the Long Meadow near the Picnic House.[6] Although picnicking is a certifiably Olmstedian use of the park, there is little sign of landscape architecture in these picnic grounds. Off-the-shelf, wooden table-bench units, some equipped with grills, are plopped down in no evident pattern. The ground around the tables is well worn. Nonetheless, picnicking and cooking out are so popular that people spread out from the established picnic/cookout grounds, through the Long Meadow and the Nethermead, and to just about any place where one can sit on the ground.

The following are excerpts from field notes taken at the lakeshore picnic area near the skating rink and the Parkside-Ocean entrance:

> By late morning, the picnicking families are arriving and setting up for a day of fun: the cooking gets going by noon and continues through the afternoon into the evening. People eat, listen to music, take walks, play ball, fish, and hang out. There are many extended family groups or groups of two and three families. Some of the picnicking groups are not families but church groups or possibly other organizations. This Saturday evening, there is a gathering on the occasion of a boy's birthday of two families from Bedford-Stuyvesant. There is a lot of cooking and eating going on. The father and another man are standing to the side of the gathering, drinking beer and talking (they seem to have had a few beers already). There's a boom box going. Women and teenagers of both sexes are sitting at and standing around the table, some organizing, some eating. The birthday boy, who is about 10, comes up to ask his father can he go somewhere with his friend (Yes but be back here soon!).
>
> People are well stocked here: they can drive in and park right next to the picnic ground, so carrying stuff in is relatively easy. In addition to coolers and bundles of food, drink, utensils, plates, cups, etc., are grills, lawn furniture; balls, bats, and gloves; and tape deck/radios (boom boxes). Varieties of popular music, including soul, Samba, disco, and jazz, emanate sometimes very loudly from the boom boxes. Cooking smoke is in the air. Litter accumulates around some of the barbecue sites as the day progresses (but from a brief tour the next morning, Sunday, it looks like the clean-up crews have covered the area).
>
> The central part of the picnic area is left open (free of picnic tables). Here men and boys are throwing balls, playing catch or having an abbreviated softball or soccer game. People from the picnicking groups also

cross the drive to play ball on the triangular grassy area near the park entrance. Girls are playing jumprope and hopscotch on the paved paths.

Hispanic visitors—mostly Puerto Rican, Dominican, and Mexican—do their picnicking mainly in two places: around the Vanderbilt Street playground on the southwest side; and in the corner of the park marked by the band shell, the entrance at Bartel–Pritchard Circle, and the Tenth Avenue gate. They patronize some of the more developed areas of the park, those with facilities like playgrounds and picnic tables, or where they can fish. Many Hispanics said they came here to socialize, to get their extended families together, especially in summer when it is hot inside people's apartments. They saw the park as a place where they could breathe freely. A Hispanic man playing cards with his sons near the Picnic House said:

> Our apartment's really small, and we spend at the park as much time as possible. Since the kids discovered the trolley, we go to a different place each weekend. We go fishing, or play soccer or baseball in this area. We also go to the playgrounds. When I come to the park I forget my problems for awhile. This area (the meadow) is my favorite because it is so big, I mean, the open space, and there are so many trees.[7]

A Mexican woman at the 11th Street playground said, "I come to the park to be in contact with nature. We live in an apartment, and contact with nature is essential to my kids." She always comes to this playground or to the Carousel, and had not visited the lake. For her, nature was the open space, and "the trees' shadows." A Guatemalan man saw the park as a meeting place to gather in after a hard workweek. He liked the "casitas" (sheltering pavilions) and "resting places" like Long Meadow, "open places—I love the view." He would like to see tennis and basketball courts distributed through the park, and places for table games.

Mexican families establish themselves in the shady grounds near the Bartel-Pritchard entrance. This is a marginal area intended as a decorative buffer zone between West Drive and the street outside the park. As at Jacob Riis Park, where Hispanics hold festive gatherings in the back-beach areas originally planted for scenic effects, the Mexican visitors here have made picnic grounds out of these fringe landscape passages.

Sports are part of the Mexican picnic scene in the Bartel-Pritchard area, particularly volleyball games. These games take place in the vicinity of the family picnics, wherever there is reasonably level ground. Games like volleyball and soccer are very hard on the grass, and only a patch of bare ground remains after a season of volleyball games in one spot. Management tries to preserve the parklike scenery of lawn and shade trees, planting trees to break up these clearings and temporarily roping off areas of worn-out grass for reseeding.

ENJOYING NATURE

Many middle-class visitors express values that reflect the Western romantic tradition of idealizing nature and wilderness, as well as present-day notions of environmentalism and civic-mindedness. These visitors tend to be aware of the artistic importance placed on the Olmsted and Vaux design and appreciative of the Woodlands Campaign. The Alliance's appeal to save Brooklyn's "last forest" is well received by this group, and the Alliance's membership is drawn largely from among them.

Most members of this constituency also place value on the park's social and recreational character. They see the natural setting as valuable in itself and conducive to social and recreational activity. These sentiments are epitomized by the words of a white professional man interviewed in the Long Meadow. He described the park as "a godsend, an island of nature, a place of respite. It's a fabulous recreation resource, a wonderful social gathering place. It's also wonderful with a dog. . . . It's a social resource in Park Slope. Just to watch the passage of the seasons in the park. I couldn't say enough of what the park means to me!"

The aesthetic ideal of countryside that the park embodies was expressed by a German woman watching a soccer game in the Long Meadow, for whom the park "reminds me of the landscape of Europe. . . . You don't see houses from inside the park, and it doesn't grow wild. Over here [the countryside] is not groomed or cultivated. This is like what I'm used to in Europe, where people can take their cars and go out into the country—there are paths and you can walk all over. All around, everywhere." She would ride out here on her bicycle from the East Village area of Manhattan to take long walks all over the park and to enjoy the cultural scene.

For many middle-class visitors enjoyment of the park consists in having a place to walk and take in the views, to be alone or walk their dogs, to run or ride a bike, or to think and relax. A few among them are devoted naturalists, some being members of the Brooklyn Bird Club. An example is a man interviewed at dusk in the Nethermead in November:

> I've logged hundreds of hours here over the years. . . . I love it in winter. I'm always here in snow and wet weather. I could never imagine living anywhere else in New York away from this park. [The Lullwater] is my favorite area. . . . Just saw a great blue heron over there. And the ravine

area . . . still natural, just the way the glacier left it. Tremendous! . . . I'm completely for wildlife; I don't like other uses at all!

These visitors represent one extreme of the usership: for them the park is important as a wildlife sanctuary rather than as a social space.

Racial and Cultural Issues

Although it can seem to occur spontaneously, the cultural diversity of uses and values described above requires active maintenance. Prospect Park would not be the successful multicultural space that it is without active work to involve the many surrounding communities in the life of the park. Community outreach has been a mainstay of the work of the Prospect Park Alliance, and racial divisions have moderated since the 1970s and early 1980s. Even so, our research in 1996 and 1997 uncovered substantial racial unease, exacerbated unwittingly by the management's own actions. Construction fencing erected in 1996 around a work zone had the effect of blocking off several paths through the park interior, contributing to the feeling on the part of some users of being excluded.

Prospect Park occupies a border zone between gentrifying neighborhoods on the north, west, and southwest; and working-class neighborhoods on the east and south. The gentrifying neighborhoods on the west and southwest sides of the park are predominantly white and Hispanic, with the prestigious blocks closest to the park being very white and middle or upper-middle class. By contrast, the neighborhoods south and east of the park are highly segregated and much poorer. For example, the 2000 census figures for the Flatbush zip code show 76 percent black, 14 percent Hispanic, and only 3 percent white. Flatbush was a middle-class, predominantly Jewish district until a period of rapid racial turnover in the 1960s.

The social environment of Prospect Park reflects the differences in class and ethnicity of the adjacent neighborhoods. Some users feel that these differences have resulted in hostility, differential maintenance, and other ramifications of racism. An older black man interviewed on a bench near the Ocean Avenue–Parkside Avenue entrance recalled the population transitions in the neighborhood since the 1960s: "It was nice then . . . everything was clean. You couldn't hardly see a black person in the park . . . only whites and Jews. . . . People in this area were very prejudiced. When minorities started to flow in, things changed. Whites moved out . . . that's when things started to go down . . . the upkeep [declined]. . . . You know when minorities move in they let things fall apart."

The same man went on to say that today everyone gets along in the park, but he feels the east side is neglected. Such allegations of discrimination in services and maintenance in the park were not uncommon among users on the east side of the park. The racial divide affects the way different groups perceive the park and its management. People will say that park resources are allocated unevenly, that the "white" or more affluent section near Prospect Park West is better cared for. Many people interviewed in the Parkside-Ocean area commented about too much brush and not enough mowing of the grass. Several people thought it was cleaner and better kept on the Park Slope side.

A few said they felt more comfortable on "their" side of the park. A man nursing a beer near the skating rink on a July afternoon said, "I like this side the best. . . . You know, my kind of people over here . . . all these black people, Spanish people . . . nice you know. I don't get the same vibe on the west side. . . . It's OK, but not for me." A man at the Willink entrance said that he had "talked to this guy who was cutting the trees and he told me to go over to the other side where the whites are. That side is much cleaner. People on this side don't take care of it. . . . I guess it's their nature. But I stay on this side where I belong."

A Trinidadian soccer player interviewed in the Nethermead emphasized the relationship between the conditions on the east side of the park and those of the adjoining neighborhoods: "This side . . . people don't appreciate [the park] like they do on the other side. Buildings people live in should be kept safe and intact you know. . . . You can't separate them. They go hand in hand, park and buildings. The fact of people being cut, AIDS, grants . . . the school system doesn't educate people to appreciate things [like the park]. Everything is a matrix." He also thought the east side needed

> more concerts to reinvigorate it . . . like, you see, they have them on the other side, but they fear the culture here. We need that too, we need a bandstand on this side where West Indian artists could play. . . . Instead we have a skating rink. Here people are of African descent, they don't use it that much—I don't say they don't use it at all. [But] we need that. There are so many steelbands.

As noted above, the fencing erected in 1996 for the Woodlands Campaign, which blocked off the major paths across the park interior, exacerbated racial unease. Some participants saw the fencing as a ploy to keep black people away from the more affluent west side of the park. A man interviewed at the Carousel that summer said, "They are fencing us out of the other side. Every light works over there, the benches are painted, the grass is cut. They're trying to segregate the park . . . barricading the routes off, fencing. Fencing . . . there are no signs."

Some said that the park imposes too many restrictions on activity and behavior. For example, one of a group of men sitting on crates near Ocean Avenue and playing cards on a June evening, said:

> They don't want you to barbecue no more. So I just play dominoes and things like that. . . . People on our block used to cook out sometimes in the park. You can't do that now without getting in trouble. I think they're making too many rules for people now. But sometimes people don't know how to act in public. I know they probably blame it on us [black people] . . . you know—"Too many black people over here." Can't let us take the park over.

Speaking of the drumming event, one regular participant said, "We did it before it was allowed. They put a curfew on us, but they let the band shell go on 'til late. You know why they do that!"[8]

Although many people of color use the west side of the park, some feel ill at ease there. A black man with in-line skates, sitting on a bench near the Meadowport Arch, had noticed our field-worker the day before. "To be honest," he told her, "you don't see that many 'sisters' out here on the weekend, at least one who is not pushing a baby carriage for a living." He said he doesn't go into the woods because someone would probably see him and scream: "It's the way things are around here. And people are afraid if you are skating alone or biking alone. The only time I feel somewhat more comfortable is if I bring my nieces and nephews—at least people see you with children and relax a bit."

A group of teenage boys relaxing on the steps of the Picnic House after bicycling agreed to be interviewed. To the question "What do you like least about the park?" one said, "White people run away! We don't run; we're not scared of them—why they scared of us? They be like, 'Uh-oh, there come some black guys.'" One boy thought that walking in the park was safer now that much of the violence had stopped. To this, another boy answered, "What do you mean?! You're all the people that's doing it!" Pointing to several of his friends in turn, he added, "Two years for rape [for you], three years robbery [for you], five years being black [for you]." Although made in jest, these comments suggest that if racial integration in the park is limited by cultural preference, it is also limited by white fear of black visitors.

Cultural Differences in Landscape Values

Another domain of information uncovered by our research relates to cultural differences in meanings relating to the park landscape. The interview data indicate significant differences between management's ideas about the landscape and those of some user groups. The differences are related to culturally bound preferences but also to people's fears for their safety.

Some people view the thickening woods as the result of neglect and indifferent maintenance. It would surely surprise them to learn that this is deliberate policy and not just neglect. A Colombian caregiver interviewed at the Vanderbilt playground complained of too much brush ("*matorrales*"): "You can't see the lake from here now. It used to be cleaner and better maintained." This woman nevertheless called the park "indispensable to my life. I would die without the park. I spend more time here than at home." She then warned the female interviewer not to venture into the interior areas across the circuit drive.

Forestry practices of cutting, pruning, and thinning trees have been contentious issues in landscape parks from the beginning (Zaitzevsky 1982; Graff 1985). Olmsted planted trees thickly, intending to thin the plantings as the trees grew up with a judicious "use of the axe." Yet efforts to thin stands of young trees have evoked furious public protests. In its 33rd Annual Report (1892), the Brooklyn Park Commission wrote that the "plantations" were becoming overcrowded in many places, with "fine trees" being crowded out by inferior, rapidly growing trees. "These trees were originally planted close to produce immediate effects, the [original designers] intending to remove many later as demanded." The commissioners admitted of a "public clamor among the ignorant that trees were being cut down and the park destroyed" when attempts had been made to thin the woods. They added that in the last two years, thinning had progressed with "public recognition that the work is necessary to the safety and beauty of the plantations" (45–46). Still, in the 1894 report, the commissioners observed the need to thin out "spindly trees blocking views." They noted a lack of color in the woods with the loss of some big trees, especially conifers, and called for a varied collection of trees, shrubs, and plants for variety and contrast.

Charles Eliot, an Olmsted disciple and professional partner of Olmsted in the 1890s, commented in 1896 on the sentiment that nature left alone produces lovelier landscapes than any "developed by intention," to which he answered "a general denial." The Middlesex Fells were more interesting than the Lynn Woods, he wrote, "because of Man and not nature. In the Fells are more pastures, more grassy glades and fields, and more variety in the height and density of the forest trees. Nature, indeed, is constantly striving to abolish even the meager existing variety, and to shut in all the paths and roads between walls of close-standing tree trunks. Thus, if the reservations are left to nature, monotony will follow" (Eliot 1999, 657).[9]

Thinning allows valuable trees to develop their "habit" without interference. Cutting makes it possible to maintain clearings and sight lines in otherwise wooded areas, which arguably enhances scenic values. Under Henry Stern's leadership in the 1980s and 1990s, the New York City Parks Department stood firmly against the cutting of trees. The authority of park foresters to exercise their judgment in pruning, thinning, and cutting was sharply restricted. One day we noticed a park worker cutting trees at the foot of Lookout Hill near Wellhouse Drive. When we asked, he said he was cutting saplings to protect a valuable collection of magnolias along the drive. He was doing this on his own accord, against the no-cutting policy. The stance against cutting and thinning allows the woods to reclaim any territory not kept open by regular lawn mowing. Many clearings in the woods and along the shores of the Lake and the Lullwater have reverted to woods in the past twenty years. This growth in the extent of woodland creates more wildlife habitat but it also restricts the uses people can make of the park. In such areas, the atmosphere comes closer to that of a wildlife sanctuary than a park.

The Prospect Park Alliance thinks of the wooded interior as a "forest" distinct from the rest of the park. One thing park officials wanted to know from the user study was how people use and evaluate this forest. Many participants, however, did not distinguish a particular forest within the park. Hispanics, in particular, said they loved the trees and shade but most seemed baffled by the questions about the forest. A participant interviewed at Parkside and Ocean, perhaps thinking of "forest" as something inclusive of social spaces, answered that the whole park was a forest. Although the confusion over the forest was partly due to a semantic discrepancy, it is also the result of different cultural ideas about what a forest is. For many white urban professionals, the term "forest" alludes to an imagined pre-Columbian wilderness. For some of the park's immigrant users, a forest may be a more inhabited place, one with trees but also elements of rural civilization: farms, orchards, villages. For them, the shady lawns around the playgrounds, the lakeshore, and the open meadows may seem as much a part of the "forest" as the thickly wooded areas.

Conclusion

In sum, the data show a variety of activities in the park, much of it distinguishable by cultural group. Afro-Caribbean music and dance was important among blacks. Picnicking and cooking out were popular among all people of color, especially Mexicans and other Hispanic groups, but were little evident among whites. White users, who in this population tended to have higher education and income, liked the ways the park symbolized community, but in practice were more likely to use the park as a place to work out or to find private refuge from the city in a natural setting. Hispanics were much more likely than either black or white users to visit the park in groups of friends and family. Users who came for personal communion with nature, landscape, or wildlife were disproportionately white. Similarly, some black men reported coming to the park to be alone with their thoughts. As one black man said, "I don't come to evaluate the park, I come to chill."

People tended to split along cultural lines over landscape values, as evidenced by comments about the reforestation of some areas. Some people favored a park landscape of open ground and trees to one of woods and brush; to them, woods belonged in the background and, while holding some symbolic importance, would generally not be an area to visit. In the same vein, such users tended to think of nature as an open ground of natural surfaces, trees, greenery, shade, breeze, and water at hand. They appreciated shelters, recreation facilities, and places to sit, and their concept of forest included such built elements. In fact, a number of Hispanics—typically interviewed in playgrounds and other developed areas of the park—were baffled by questions that implied that the park's "forest" was somewhere else: for them the shade trees and open ground were forest too. Management saw the interior woodlands as a distinct forest area more precious than the ordinary recreational landscapes of playgrounds, fields, and roadway.

As the leisure research literature has shown, blacks may experience racial discrimination in urban parks and other public settings—even where people of color are a majority. Management actions not grounded in ethnographic knowledge can appear discriminatory to some users. For example, the woodlands construction fencing disrupted the communication between different parts of the park. As a consequence, some users on the poorer, east side of the park felt as though they were being fenced out of the more affluent west side. Similarly, the reforestation policy in marginal areas has not taken into account the culturally based preferences of many people of color for clear, open ground, leaving some to attribute the condition to official neglect.

Prospect Park is successful in attracting such a great variety of people and activity for several reasons. First is a basic enticement intrinsic to a conveniently located, naturalistic, open ground enhanced with recreational amenities. The spatial pattern of open green spaces defined and separated by vegetation and topographic features, and the presence of water in some areas, provides many places for people to gather. This layout allows most visitors to find the combination of public and private experience that feels right to them. As we have seen, many locations become identified with one or another group, some defined by ethnicity and others more by activity.

An important variable in the distribution of such places is the presence of built features either intended or adapted for recreational use. Many of the socially active places in Prospect Park combine natural features of ground, vegetation, and sometimes water with built features. The location of the Hispanic picnic areas on the southwest perimeter close to the park drive provides easy access and high visibility. Nearby are two playgrounds, which are major attractions for families. People feel safe here. Some of the more intensively used spots are paved ground no longer open to vehicular traffic—a closed park entrance, an old carriage turnout—where children roller-skate, play hopscotch, and so on. The African drumming site is also at a visible, accessible location adjacent to the park drive and close to the prominent Parkside-Ocean entrance.

A second important factor is that Prospect Park is the only large-scale recreation ground in a borough of nearly 3 million, few of whom have their own backyards. That alone implies a demand sufficient to produce considerable activity in the available open space, whatever its condition. Third, since the administrator's office was established in 1980, management has improved and renewed the physical fabric of both landscape and facilities and has worked tirelessly to attract groups into the park and build park constituencies among Brooklyn residents. Although attendance has at least doubled over that time, the park had three times as many annual visitors a century ago. We guess that the smaller attendance today is as much the product of changed recreational patterns, greater affluence, and easier access to mountains and beaches beyond the city limits as it is a response to the condition of the park. Although the park is in much better shape than it was in 1980, it falls considerably short of the offerings, amenities, and standard of maintenance of a century ago.

Knowledge of class and cultural differences provides a basis for making the park environment as responsive as possible to its users' values. The regeneration program of the Prospect Park Alliance conceives of the park first as an ecological system and artistic creation in need of conservation and restoration, and second as a city park. Prospect Park may be a valuable natural resource, but contemporary city residents have access to more satisfying examples of natural environments a bit farther afield. What really attracts people to this city park is its vibrant cultural life in a naturalistic setting. While it is a good thing to take care of the natural and artistic resources, it is far better to do so from a culturally informed basis.

The original Olmsted and Vaux design provided certain "places of congregation," as they put it, designed to accommodate throngs of visitors. The present restoration program devotes relatively large resources to wooded areas that few people visit, and does little for the places of congregation. A management program that saw the park as first a social space would seek changes that improved the fit between uses and resources. The data generated by the user study are a source of important information on uses, users, and cultural values. If, as the study shows, thousands of visitors enjoy picnicking under shade trees and in proximity to playgrounds, water, and the roadway, why not put landscape architects to work on better accommodating those areas to such activity? If the African American/West Indian community sustains an extraordinary drumming and dancing event over many years, why not work to redesign the grounds to both accommodate and celebrate the practice? If it can be plainly observed that people enjoy wading in the stream, then dense streambank planting and protective fencing should not be the only way of restoring the streambed. Social and cultural data need not be the only planning criteria, but they ought to be included with and complementary to artistic and ecological imperatives.[10]

Notes

1. In 2004 restoration has reached the shoreline edge of the peninsula meadow. However, the general atmosphere described in these paragraphs remains.

2. The Concert Grove was first upstaged by another musical performance facility, the music pagoda, constructed in the 1880s. The Concert Grove's spatial integrity was damaged in 1960 by the construction of a skating rink that preempted a portion of its territory. The Lookout was compromised by a lack of maintenance that allowed the view to become obscured by trees.

3. In a letter to the authors dated September 2004, Ms. Thomas wrote:
The Alliance does not make any management decisions about the Park. (I work for Parks and the Alliance.) The decision to restore the natural areas of the Park was NYC and was funded by NYC. The Alliance supported the efforts by providing funding for a natural resource crew and for educational programs. It is important, in fact essential, to understand that the private side of the partnership was involved in supporting the public side. I had been employed by Parks, as the Administrator, since 1980. It was as Administrator, that the decision to allow barbecuing and to develop picnic zones with tables was made. The development of privileges for dog owners, restoration of every playground and restoration of the forest was all made by government.

4. In reviewing this chapter in September 2004, the Administrator, Tupper Thomas, wrote that
the main issue for me is how much time has elapsed between your study and now—about 10 years. We have developed, since 1996, a strong Community Committee; we have spent millions on the east side of the Park; we have developed even more barbecue and picnic areas and usership has continued to grow, but remains very diverse. We have found that the Mexican/South American populations have been replaced by the next group of new immigrants—Russians and Pakistani.

5. Charles Price-Reavis is now an assistant professor of anthropology at the University of North Carolina, Chapel Hill.

6. No picnicking services or supplies are offered. The main floor is rented out for private functions, and the basement is used for administrative offices. The building does have public restrooms, a pay phone, and candy and beverage vending machines.

7. The "trolley" is a bus designed to resemble an old-fashioned open streetcar that makes a circuit around the park and other destinations in the vicinity.

8. The Bandshell is on the west side of the park.

9. The Middlesex Fells Reservation and the Lynn Woods Reservation are public lands near Boston associated with Charles Eliot and the Metropolitan Park Commis-

sion. Each is approximately 2,000 acres. Middlesex Fells was one of the first acquisitions of the Park Commission in the 1890s; Lynn Woods was established by the city of Lynn with advisement from Frederick Law Olmsted in 1878 (Cushing 1988).

10. Administrator Thomas writes, in response to this chapter,

There are often times that government has to make decisions about a park, which are for the greater good—like maintaining the eastern seaboard flyway, restoring the landmarks design, and improving the Park's buildings. Even in the 1980s, before the creation of the very diverse Community Committee, there was an advisory committee of five community boards, local officials, and about 10 groups who reviewed all plans. Every capital project also had enormous involvement of specific user groups.

Chapter 4

The Ellis Island Bridge Proposal
Cultural Values, Park Access, and Economics

Introduction

In 1994 the Public Space Research Group was asked by the National Park Service to find out what local residents thought about building a bridge from Liberty State Park in New Jersey to Ellis Island. Ellis Island was the federal immigration station for the Port of New York from 1892 to 1954. More than 12 million immigrants were processed there, and over 40 percent of all U.S. citizens can trace their ancestry to those who came through this facility. In its early years, when the greatest number of immigrants arrived, Ellis Island represented an "open door" policy to the growing cultural diversity in the United States. After the passage of restrictive immigration laws in the 1920s, however, it became a place of assembly and often detainment. Immigrants were required to pass a series of medical and legal examinations before they were allowed to enter, and those who could not pass these tests were deported. In 1954 Ellis Island was closed, and it remained abandoned until 1965 when President Lyndon Johnson added it to the Statue of Liberty National Monument. Restoration began in 1983 and the Ellis Island Museum opened in 1990.

The restoration project included the construction of a bridge across the 400 yards of water between Ellis Island and Liberty State Park, on the Jersey City mainland. The bridge allowed construction vehicles, personnel, and equipment to be driven right onto the island thereby avoiding a costly transfer to water-based transport. The bridge was and remains in full view of everyone using Liberty State Park's bayfront promenade, Liberty Walk, and park users began inquiring at the security booth whether they could walk over to Ellis Island. The answer was no; the bridge had been built only to service the Ellis Island restoration project and was not intended to invade the Circle Line ferry's passenger market. However, the public clamor for access to the bridge led to a congressional appropriation for a permanent bridge to allow limited vehicular and unrestricted pedestrian access to Ellis Island. The authors' study was part of a federal environmental impact statement on the proposed bridge project. Funds for the permanent bridge were rescinded some years later; the supposedly temporary bridge remains in use for official business and off-limits to the public.

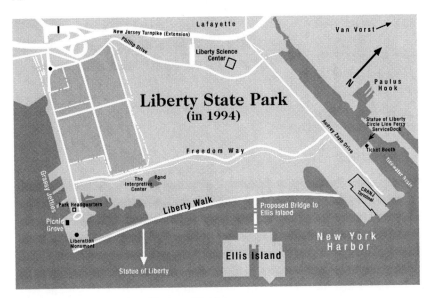

Map 4.1. Liberty State Park and Proposed Bridge

Ellis Island can only be reached by a ferry that services both the Statue of Liberty and the museum site. The majority of visitors leave from the ferry landing at Battery Park, located at the tip of Manhattan in New York City, but a New Jersey ferry from the marina near Liberty State Park also provides limited service. There has always been considerable contention about who "owns" Ellis Island. New York and New Jersey have joint jurisdiction of the island and its surrounding waters and share the revenue. However, for most visitors, Ellis Island is a New York–associated tourist event.

In order to make Ellis Island more accessible to New Jersey residents, Senator Frank Lautenberg of New Jersey introduced a bill in Congress that would allocate $15 million to construct a bridge from Liberty State Park to the National Park Service site. He argued that the people of New Jersey needed better access to the island that lay only four hundred meters off their shoreline and that it would promote tourism and encourage the people of New Jersey to visit this national treasure. Further, it would complement the developing of Liberty State Park on the Jersey City shore.

Historic preservation groups in New Jersey and New York, the Circle Line ferry company that provides access to Ellis Island, and the tourist office and mayor's office in New York City had already organized the opposition by holding numerous public meetings about the dire economic consequences of building a bridge. The *New York Times* ran a series of scathing commentaries about New Jersey's inability to manage the barrage of tourists expected to arrive by

THE ELLIS ISLAND BRIDGE PROPOSAL

this new route. The Municipal Arts Society and the National Park Service worried that a bridge would destroy the experience of visiting this immigration entry point that had always been approached by water. No one, though, had actually spoken to the local Jersey City residents and users of the parks and surrounding area that would be most affected.

We were hired by the NPS to collect the missing commentary. The pedestrian bridge and three alternatives—a subsidized ferry, elevated rail, and possible tunnel—were to be considered from an ethnographic perspective, that is, from the point of view of culturally appropriate and relevant populations whose opinions had not yet been elicited. We were to contact local users of Battery Park and Liberty State Park; local providers of services at Battery Park and Liberty State Park, including vendors and small-scale tourist services; residents of the Jersey City neighborhoods adjacent to Liberty State Park; and special populations such as children, the elderly, and the physically challenged. The "traditional cultural groups"—those people whose families entered through Ellis Island or who are themselves immigrants with identities and aspirations symbolically connected to Ellis Island—were surveyed by a questionnaire, while other interested community members, particularly advocates for preservation and no bridge, had already participated in the earlier public hearings.

Methodology

Our study began by focusing on "constituency groups," that is, people who share cultural beliefs and values and are likely to be affected in the same way by the proposed bridge and its alternatives. Our goals were to develop a list of groups that would reflect the diversity of users and residents and to provide a guide for interviewing people in the parks. We would then modify the rapid ethnographic assessment procedures (REAP), discussed in Chapter 8, to collect the needed information. To see the constituency groups we identified, refer to Table 4.1 on the next page.

A number of REAP methods were used. Behavioral maps of Battery Park and Liberty State Park sampled both weekends and weekdays from 8 a.m. until 8 p.m. Physical traces maps of Battery Park and Liberty State Park that recorded the presence of liquor bottles, needles, trash, clothing, and erosion were collected early each morning. We completed a four-hour transect walk (a guided tour of important places by a local; see Chapter 8) in Battery Park with a can collector and two lengthy transect walks in Liberty State Park with a local librarian and an active community member.

Individual interviews were completed in Spanish, Russian, or English, depending on the preference of the interviewee. A version of this interview was also used in the surrounding neighborhoods of Paulus Hook, Van Vorst, and

Table 4.1. Ellis Island Bridge Constituency Groups

Battery Park	Liberty State Park	Jersey City Neighborhoods
Homeless residents	Homeless residents	Local residents
Vendors	Vendors	Vendors
Local businesses (concessions)	Local businesses (concessions)	Local businesses (bars, barbershop, bodegas)
Transportation services	Transportation services	Parking garages, buses, car service
Street performers	Cultural and scientific institutions	Cultural and religious institutions
Park users	Park users	Park and street users
NPS park rangers	Park rangers	Local school
Tourists	Tourists	Tourists
Battery Park Conservancy	Friends of Liberty State Park	Social services and nonprofit organizations (senior centers, after-school programs)

Lafayette. We collected 41 interviews in Battery Park and 76 interviews in Liberty State Park, for a total of 117 individual interviews.

Expert interviews were collected from those people who were identified as having special expertise to comment on the Ellis Island bridge and alternatives, such as the head of the vendors' association, neighborhood association presidents, the head of the planning board, teachers in local schools, pastors/ministers of local churches, principals of local schools, and representatives from the Liberty State Park and Battery Park administrations. We collected four expert interviews in Battery Park, and five expert interviews in Liberty State Park. We also collected five expert interviews in the surrounding neighborhoods.

Impromptu group interviews occurred where people gathered outside of public places or at special meetings set up with church groups to discuss the alternatives. They were open-ended and included any community members who were interested in joining our discussion group. We held group interviews in the neighborhood barbershop, on a street corner where people were waiting for the bus, in the local public library, on stoops, on porches, in bars and neighborhood restaurants, in a Catholic convent, and in front of churches on Sunday morning. We collected 113 impromptu interviews in the neighborhoods.

Focus groups were set up with those constituencies that we thought were particularly important in terms of understanding the potential impact of the bridge. As opposed to the large, open group interviews, the focus groups consisted of 6 to 10 individuals selected to represent especially vulnerable populations such as schoolchildren, seniors groups, and physically challenged groups. We obtained permission to run the focus groups through the agency that organized the groups' activities. In the case of the children's focus group, the focus group became one of the activity choices designated by the head of the organization. The discussions, which were conducted either in English or in Spanish, were directed by a facilitator, and were tape-recorded. Two focus groups with children (21), five with seniors (32), one with a Spanish-speaking church group (17), and one with an African American Baptist church (18) were collected, for a total of 88 persons consulted in focus groups. This brought the total number of people consulted to 318. The opinions collected about the alternatives were analyzed for content and presented as lists of arguments for and against the bridge, and a value orientations analysis summarized the various positions held across the subgroups.

Table 4.2 reviews the methods used, how much time was spent on each, the kind of information that was produced, and what was learned.

Table 4.2. Ellis Island: Methods, Data, Duration, Products, and What Can Be Learned

Method	Data	Duration	Product	What Can Be Learned
Behavioral Mapping	Time/space maps of sites, field notes	2 days	Description of daily activities on-site	Identified daytime activities that would be affected
Physical Traces Mapping	Map of trash and clothing left in parks	1 day	Description of physical condition of site	Identified nighttime activities that would be affected
Transect Walks	Transcribed interviews and participant's map of site, field notes	4 days	Description of site from community member's point of view	Community-centered understanding of the site and local meanings
Individual Interviews	Interview sheets, field notes	10 days	Description of responses of the constituency groups	Community and user responses to the proposed bridge
Expert Interviews	In-depth interview transcriptions	5 days	Description of positions of local institutions and community leaders	Community leaders' responses to the proposed bridge

Table 4.2. (*continued*)

Method	Data	Duration	Product	What Can Be Learned
Group Interviews	Field notes, video or tape recording	5 days	Description of various community groups and their responses to bridge	Involved the neighborhood and church groups in the planning process
Focus Groups	Field notes, video or tape recording	2 days	Description of issues that emerge in small group discussion	Enabled the development of a typology of value orientations

Findings

Battery Park

PHYSICAL SETTING

Battery Park is one of approximately 1500 parks, playgrounds, and other public spaces in the jurisdiction of the New York City Parks and Recreation Department. The park covers almost 23 (22.98) acres of land in a tear-shaped tract that stretches between State Street, Battery Place, and New York Harbor. Winding pathways lined with wooden benches divide the park into spaces of flat grassy areas dotted by shade trees, picnic tables, and small enclosures that protect a variety of monuments. A black iron fence outlines the park's perimeter and subareas. Many informal paths throughout the park have been created by park users through the erosion of the original surfaces. The largest section of lawn, the Great Lawn, is circumscribed by trees and fenced off, and park users are not permitted in this area.

The principal entrance to the park is at Bowling Green, the intersection of State Street and Battery Place, by the entrance to the #4/#5 subway station. To the east of this entrance is a city tour-bus stop and beyond the bus stop is a taxicab line. West of the entrance is an area for other tour buses to deposit and collect tourists. Eisenhower Mall is a tree-lined walkway flanking the Hope Garden, which links the park's main entrance with Castle Clinton. The surrounding walkways are lined with benches facing each other. West of lower Eisenhower Mall is a restroom, which was temporarily closed for construction. Castle Clinton, a round, red stone fortress built in the 1800s, is a national monument under federal jurisdiction. The enclosed structure contains its own historical exhibitions, offices, gift shop, and the Circle Line ticket kiosk; people

THE ELLIS ISLAND BRIDGE PROPOSAL

Figure 4.1. Circle Line ferry from Battery Park to Ellis Island

buy ferry tickets here for the Statue of Liberty and Ellis Island. It is the major site of tourist activity in the park. Castle Clinton has entrances onto both Eisenhower Mall and the harbor's Admiral Dewey Promenade, both of which are lined with pushcart vendors. The gun emplacement apertures of Castle Clinton are used as shelter by couples and young tourists in the rain, and by homeless individuals at night. The entire structure is enclosed by a guardrail, which serves to control lines of tourists waiting to board the Liberty and Ellis Island ferries.

Admiral Dewey Promenade is a curved esplanade that connects Castle Clinton to the harbor and outlines the southern edge of the park. Along the eastern edge of the promenade is a war memorial plaza, an abandoned kiosk, another outdoor restaurant, and docked boats in the harbor (figure 4.2). Between Castle Clinton and the south end of the park are located the ferry landing and various harbor pier posts. This part of the park is elevated above the promenade and is lined with pay-binoculars and benches. Pushcart vendors and street performers concentrate their work along the promenade and on the elevated area. The promenade is just wide enough for performances to take place in front of the long lines of tourists, for the vendors to sell T-shirts, and for dump trucks and patrol cars to be able to pass by (figure 4.3).

Figure 4.2. Battery Park landscape with Castle Clinton in the background

Figure 4.3. Caricatures for sale, Battery Park

In the southeast sector of the park there is a functioning restroom and a city workers' changing room located at the corner of the enclosed playground area. The northwest corner of the playground is an abandoned horseshoe pitching area. Iron stakes remain, surrounded by weeds. This area has one entrance, locked by a chain; nevertheless, homeless individuals frequently occupy these benches. Beyond the playground and across the street are concession stands, several pushcart vendors, the entrance to the #1/#9 subway line, and the entrance to the Staten Island ferry terminal. In front of the concession stands are portable picnic tables and a city bus stop.

SOCIAL SETTING

Visitors to the park include tourists, the Wall Street lunchtime crowd, and New York and New Jersey residents who have come to enjoy the park. There are active users engaged in such activities as walking, shopping, biking, in-line skating, and fishing, and passive users who are resting and observing.

Tourists can be found throughout the park, although the majority tend to convene near the ferry landing, souvenir pushcart area, and tour-bus area. Their activities consist of walking, eating, resting, observing, and shopping. Streams of tourists exiting the ferry generally flow through the wide pathway connecting the promenade to the great lawn area. Workers on their lunch hour tend to sit both in the sun and in the shade of Eisenhower Mall, around the great lawn, and in the picnic areas. They can be found reading books or magazines, eating a brown-bag lunch, listening to music with headphones, talking in pairs or in a group, or lunching at the east side restaurant. Other users sit facing the water or city while resting, passing time, or waiting for a friend. Students from a nearby language institute use the park to meet each other between sessions, or to practice English with strangers. Other visitors to the park engage in a variety of activities and make use of different areas of the park. Fishermen tend to gather at the end of the harbor. Bikers and in-line skaters use the length of the promenade. Sunbathers can be found along the edges of the promenade, where there is the least shade.

Several different groups of park and recreation workers serve Battery Park. National Park Service rangers are generally inside or at the door of Castle Clinton directing tourists, or giving guided tours in the park and surrounding neighborhood. City Parks and Recreation employees work throughout, maintaining the park. Several city employees regularly lunch in the playground area at the picnic tables. A police car patrols the park for security, and often city park officers are seen talking with homeless individuals or illegal street vendors.

There are many kinds of vendors in Battery Park. Three vending companies—two that concentrate on food and one on souvenirs—have regular pushcarts in the park. Another pushcart company employs the majority of immigrants operating pushcarts near the entrance to the park, Castle Clinton, and the nearby tour-bus stop. The third company occupies territory near the bus stop as well as near the ferry landing. Independent vendors are spread out between the two ends of the park. Two days a week there are large flea markets along State Street and Bowling Green. The vendors who participate in these flea markets work out of wooden stalls, leasing space from a private company for the entire day. The majority of vendors who participate in these open-air flea markets are also immigrants.

Circle Line ferry workers are located in four areas within the park. The main ticket kiosk is inside Castle Clinton. Lines of people trying to buy tickets stretch out the door of the fort and during high tourist season wrap around the building. Another minikiosk stands in the park outside of the Castle, although there is often no Circle Line employee there. Several men in charge of docking the boats stand near the ramps when the ferry is moored. Finally, several young men and women take tickets and manage the crowds in line along the promenade. (After 9/11 a metal detector station was added to this already chaotic scene.)

Street performers, most of whom are immigrants, position themselves on the promenade where people wait to board the ferries. Performers who arrive later in the day claim sites on the far eastern side of the park near the marine flagpole. Street performer activities include acrobatic tumbling; pantomime; steel drumming; horn, accordion, and guitar playing; singing; and exhibiting wildlife curiosities. Since performers spend many hours a day in the park, they also interact with tourists, visitors, and other performers and patronize pushcart vendors.

There are several citizens in the park who collect discarded cans and bottles to redeem the nickel deposits. Their opportunities are increased on the many days when the park's trash bins, which are positioned throughout the park, are not emptied. Some can collectors also have informal agreements with vendors and park workers who save cans specifically for them.

A large number of homeless and unoccupied people reside in the park. Groups of unoccupied men and women gather in the eastern part of the park. The stone slabs of the war memorial offer privacy to a person sleeping on a bench, and patches of healthy grass, bathrooms, and a running fountain, all of which can be found in the park, are precious resources to the homeless residents. Homeless individuals are found throughout the park, including in areas deemed off-limits to park users. By evening the homeless residents outnumber other park users. A service center for homeless individuals is located underneath the Staten Island ferry terminal, and there is a soup kitchen in the surrounding neighborhood.

THE ELLIS ISLAND BRIDGE PROPOSAL

CONSTITUENCY GROUP FINDINGS

The number of constituency groups consulted was 15. Of those, 12 groups have work-related reasons for being in the park, two groups visit for recreational purposes, and one group resides in the park. A total of 41 people were consulted, 35 of which are men and 6 women. Half are native-born Americans and identify themselves as African American (12) or white (8); the other half are immigrants from Africa, Europe, South America, the Caribbean, and the Middle East. There was one man interviewed who held a job on Ellis Island who was neither a permanent immigrant nor a citizen of the United States.

Of those consulted, 28 individuals were working in the park when encountered. The following is a summary of the different constituencies of workers and their responses to questions about access to Ellis Island.

Most of the street performers knew about the proposed bridge and viewed it in direct relationship to their work. Some were concerned that a bridge from Liberty State Park to Ellis Island might reduce the number of tourists coming to Battery Park, and thus have an adverse impact on their ability to make an adequate living. They thought the vendors would suffer as well, and felt that the money should be spent differently. They also felt that the alternatives are a "waste of tax money . . . leave it like this . . . there's a breeze and a boat ride for $6.00; it's beautiful." (It is now $7.50 at time of publication.)

The attitude of the employed pushcart food vendors toward the proposed new bridge access to Ellis Island was positive: "The bridge is good—more people will go to it—this means more customers and more business. It's a good thing. People can go there on their own and relax there. The population is going up, and there has to be something to make it easier to visit the island. The bridge can help. People now have to wait a long time in line. . . . America is about choice. People should have a choice." Independent pushcart vendors also thought that the bridge was good because, as one said, "potentially it could bring more people to Battery Park, if New Jersey tourists took the ferry over to New York from Ellis Island." Informal vendors wanted to know more but thought the bridge was "a good thing. . . . I can go with my children" or "it's nice to walk. It could not be a far walk." Flea market vendors also imagined themselves enjoying the visit walking over the bridge.

A can collector, on the other hand, said that he felt that the bridge was "a waste of taxpayers' money. . . . There are more people now on welfare. . . . People need houses, jobs, and to get the kids off the streets." He felt that the bridge was going to affect the people who work for the ferry and people like himself who pick up cans: "You can't pick up cans on federal property." And the three city employees agreed that "it is better to go on the boat." They feel that a bridge or any of the alternatives would affect their jobs. One participant speaking for the group said: "A lot of people come from New Jersey State to New York. . . .

It would take a whole lot of money out of the economy and then that would probably cut me out of a job. . . . I am telling you I have kids. . . . I need all the money I can get." Park rangers, however, thought that "there's no difference between a boat or a bridge, although the ferry is part of the visit. . . . Improvements can be made with better coordination with Circle Line. More dialogue is needed. . . . Less people would come here if Circle Line were not here." The other ranger explained: "It won't affect this park at all . . . for tourists coming to New York they are staying in Manhattan and this is convenient for tourists. . . . The bridge over there would be a tremendous benefit to New Jersey. You can park over there." They agreed, however, that arriving by water is key to the Ellis Island experience.

Finally, the ferry workers were concerned that "there would be some loss of passengers to us, but the real problem is not that, it is the ticketing. Some people will be paying nothing to go to Ellis Island, but then will want to go on to Liberty Island. The problem will be to distinguish the people who want to go to which island. . . . How much loss of revenue is unknown, but definitely there would be a loss of some revenue."

The remaining groups were recreation-related constituencies—11 men and women were interviewed while recreating in Battery Park. Six were identified as passive park users (sitting, reading, waiting, sunbathing), and five as active park users (biking, in-line skating, and fishing). Of these, 50 percent thought a bridge might increase their chances of one day visiting the island, and that it is a good idea: "A footbridge doesn't sound too bad. . . . I wouldn't like to see cars." Most do not like the idea of the other alternatives because they would not be ecologically sound or would be too commercialized. "There's a lot of swamps, it might affect [the environment]." One person thought that a tunnel connection would preserve the island status of Ellis Island.

Some were strongly opposed to the bridge: "I prefer the ferry . . . leave it alone . . . you'll end up spending unnecessary money . . . with the money . . . give it to the homeless." Another said that "it would spoil the beauty of Ellis. . . . Some things are better left alone. . . . Ellis Island doesn't have a history of a footbridge . . . leave it alone." Others were strongly in favor: "It is a good thing. It will help make life better, and bring more work." "New Jersey should have it . . . if they have all the concessions stands there, they should benefit from it, and it would mean that you would be giving people more of a choice." The fishermen were concerned about environmental damage from the proposed alternatives: "How much environmental damage it will do to the river! They'll just build it like they always do. . . . The fishermen won't know nothing about it until it is all over."

Of the people who reside in the park, one middle-aged African American man and one elderly European American man were consulted as they sat on

THE ELLIS ISLAND BRIDGE PROPOSAL

benches within the center area of the park. One was not particularly interested in being interviewed and did not see how the bridge or any other alternative would change the park. Another was strongly opposed to the bridge. He considered the bridge and the alternatives a waste of money. "Who's going to pay for this bridge construction?" was his question.

Overall, the constituency group findings ranged from strongly stated concerns about the negative impact of the bridge or other alternatives to strongly felt statements about the positive increase in access and choice that would occur with the building of a bridge (tunnel, elevated rail). The subsidized ferry option did not produce clear responses from the Battery Park constituency group, possibly because most of the people interviewed were not concerned with visiting Ellis Island, and therefore did not focus on the issue of the cost of the ferry ride. Surprisingly, the vendors who potentially could be negatively affected by increased New Jersey access to Ellis Island were not very concerned about it, and in fact were positive about the idea of the bridge and the alternatives. The street performers, however, who are also dependent on the number of tourists for their livelihoods, clearly articulated concerns about the impact on their work.

Less surprising was that service managers, city employees, park employees, the ferry representative, the can collector, and the tour bus driver were concerned that the proposed increased access from New Jersey would have negative consequences for their work, profits, or work environment. These people have a vested interest in the success and profitability of Battery Park, while the vendors and concession workers do not perceive themselves as closely tied to the park.

VALUE ORIENTATIONS ANALYSIS

There are a number of distinct value orientations that people use when thinking about the impact of the proposed bridge and access alternatives. These value orientations do not necessarily predict which of the alternatives or attitudes (positive or negative) interviewees consider most appropriate, but they do provide insight into the way that participants think about the Ellis Island access problem. These value orientations represent the types of concerns that need to be addressed in the social impact assessment process.

The list of value orientations (Table 4.3) is a distillation of all the responses gathered in Battery Park about access alternatives to Ellis Island. They are presented in the order of the number of participants who mentioned them during the individual interviews (from most mentioned to least mentioned). Most participants discussed more than one orientation; the total number of responses (116) is therefore greater than the number of respondents (41). Not surprisingly, the most prevalent value orientation in Battery Park is economic (23), reflect-

Table 4.3. Value Orientations at Battery Park

Value Orientation	Examples	Number of Responses
Economic	"good for business" "bad for the ferry"	23
Access	"will allow more people access to Ellis Island"	13
Social Priorities	"the money should be spent on the homeless" "we should be helping children stay off drugs"	10
Choice	"you will lose the ferry option" "it's democratic, people can choose to walk or ride"	9
Health and Recreation	"it's fun for children to walk" "it's healthy for people to walk"	9
Political	"it's part of the New York and New Jersey conflict" "it's a [political] hot potato"	8
Personal	"I do not want to walk"	8
Aesthetic	"nice view from the bridge" "how will it look?"	6
Park Quality	"it will improve Ellis Island"	6
New Technology	"it is progress" "it is modern"	5
Safety and Comfort	"People feel safer on a bridge than in a tunnel"	4
Education	"people could learn something"	4
Ecological	"it is a swamp" [and should be protected]	2
No Impact	"it is not going to make any difference at all"	9

ing the concerns of the large number of workers interviewed (28). But there are also a large number of responses that relate to access, choice, and social priorities (42), involving evaluations of the larger sociopolitical implications of this decision for the majority of people. And only eight individuals were concerned solely with the impact of the proposed changes on themselves.

Liberty State Park
PHYSICAL SETTING

Liberty State Park occupies 1,122 acres of land and tideland along Upper New York Bay in Jersey City, New Jersey. The site was a vast railroad yard through most of the twentieth century. By the 1960s, all passenger rail and freight operations on the site had been abandoned. The state of New Jersey acquired the site and has been gradually transforming it into a public park. The first phase of Liberty State Park opened in June 1976, in time for the national Bicentennial observances. The area developed for park use so far comprises approximately 300 acres, mostly at the southern and northern edges of the park.

The southern area, which was the first part of the park to be constructed and opened, and is the most intensively used area in Liberty State Park, includes grass-covered fields, a public boat launch, walkways along the waterfront, spacious parking lots, and the park headquarters, which contains a food concession stand, restrooms, and visitor information. Adjacent to the headquarters is the picnic grove. Also in the southern area, but set off from the rest of the park, are a swimming pool and tennis courts. The swimming pool is well used, while the tennis courts are in disrepair and often locked.

The northern area has three centers of activity widely separated by flat, mostly treeless fields (figure 4.4). Two of the activity centers are major developments of recent years: the Liberty Science Center and the restored headhouse and concourse of the Central Railroad of New Jersey (CRRNJ) ferry terminal, where passengers once boarded ferries for New York. The soaring roofs of the waiting room and concourse now shelter exhibits, lectures, and various special events, and a new weekend ferry service again connects the terminal with downtown New York. Passengers disembarking here can take a jitney up Audrey Zapp Drive to the Liberty Science Center, a "hands-on" science museum opened in 1992 that attracts busloads of schoolchildren and other visitors from all over the metropolitan area. The science center was built at the western edge of the park, far removed from the CRRNJ Terminal.

The third concentration of activity in the northern area is the Statue of Liberty and Ellis Island passenger ferry dock, which is located along the Tidewater Basin. Visitors to the two national monuments can park in the newly constructed parking lot across Audrey Zapp Drive, west of the CRRNJ Terminal train shed, and walk across the street to the ferry dock, from which the ferries come and go at roughly 45-minute intervals. Next to the dock are a ticket stand operated by the ferry company, a film and souvenir stand, several picnic tables, sheltered waiting areas, several refreshment vendor carts, and public lavatories.

The northern and southern areas of Liberty State Park are connected by two corridors through the partly undeveloped interior land. Freedom Way, a divided parkway with a parallel bicycle and jogging path, gives access to the Interpretive Center, an educational visitor center that houses exhibits on the park's salt marsh habitat and is the gateway to the park's designated "natural area," 60 acres of salt marsh and protected upland habitat. The other connecting corridor is the 1-1/3-mile-long Liberty Walk, a newly built promenade along the water's edge that offers sweeping views of Upper New York Bay with Ellis and Liberty Islands as central features of the panorama. The present Ellis Island

Figure 4.4. A meadow in the northern sector of Liberty State Park

bridge interrupts the flow of Liberty Walk at a point about one-third of the way from its northern end near the CRRNJ Terminal to its southern end near park headquarters.

SOCIAL SETTING

The most popular area includes the perimeter walkways around the Liberation Monument, east of park headquarters, and the section of Liberty Walk that continues from this field on a trestle across the south cove. This area is within easy walking distance of two large parking lots and offers spectacular views of the bay, the Statue of Liberty, and the New York skyline. Liberty Walk itself has numerous benches where people rest and enjoy the views and the breezes. From the southern terminus of Liberty Walk, pedestrians follow a path along the shoreline past the Liberation Monument field, around the picnic grove, then past a series of jetties that extend into the bay on a southerly axis. Sheltered benches are found at frequent intervals all along this path.

The grassy jetties are the favorite places in Liberty State Park for sunbathing. People carry folding chairs only a few steps from their parked cars out onto one of the jetties and sit or lie there alone or in pairs or family groups. The path

THE ELLIS ISLAND BRIDGE PROPOSAL

between the parking lots and the jetties is the resort of elderly people, often men, who fraternize on the sheltered benches or stroll along the path under the canopy of plane trees in pairs or small groups. Toward evening, some people simply park their cars along the waterside of the parking lots and gaze at the water without getting out.

The picnic grove is used by families, organized groups, and some individuals. People can buy take-out food at the stand in the park headquarters, but most seem to bring picnics, and some cook on outdoor grills. The picnic grove and the adjacent field containing the Liberation Monument is popular with people supervising children and is also popular with kite flyers. Family groups also stroll along the walkways at either side of the field out to and along Liberty Walk, admiring the view. Some continue along Liberty Walk, but most turn around and come back. Some people take advantage of the pay-binoculars located along the seaboard railing.

Most people do not walk the length of Liberty Walk: it is long and the middle stretch offers little variety. Many of those who do cover the whole distance of Liberty Walk on foot do it for exercise. Some jog, usually alone; others do forms of exercise walking, often in pairs. Some go from one end of Liberty Walk to the other and back; others do a circuit that includes Liberty Walk and Freedom Way. Liberty Walk is also popular with cyclists and in-line skaters. The north and south coves are popular with people fishing, especially in the early morning and in the evening. The fishermen stay several hours at a time, and many know one another. One fisherman, for example, leaves his lines for 10 or 15 minutes to walk along the walkway, stopping to talk with other fishermen, before returning to his chosen station.

In the northern area of the park, there is continual activity during the day at the Ellis and Liberty Island Circle Line ferry dock. Visitors to the national monuments who come by car park in the pay lot opposite the ferry dock. After purchasing their tickets, the visitors often have a while to wait before boarding the boat. Some browse through the souvenir stand or among the pushcarts, while others sit down on the benches or at the picnic tables. Visitors to the national monuments rarely go anywhere in Liberty State Park other than the ferry service dock area and parking lot.

A few people play catch or sunbathe between the basin and Audrey Zapp Drive. Some dog owners walk their dogs over from Van Vorst Park to let them run on the open field. East Indians use the field farther east as a gathering place, principally on Friday evenings. One recent Friday evening, women in saris and children were sitting in a circle, singing, while men talked in clusters—about 50 people altogether.

The brick plaza next to the CRRNJ Terminal is used occasionally in the daytime for ceremonies, such as Flag Day observances by the Jersey City Fire Department. In summer, Jersey City sponsors Sunday afternoon jazz concerts here. On sunny evenings, people may drive down to the plaza, parking in the free lot next to the ferry terminal, to watch the sunset. The CRRNJ Terminal is lightly used on weekdays by people visiting the historical exhibits in the old waiting room, using the lavatories, or just looking at the building itself. However, on some weekends the terminal is used for special events like ethnic festivals or collectors' shows, which may attract thousands of people.

CONSTITUENCY GROUP FINDINGS

We consulted a total of 76 individuals, who are distributed among 12 different constituency groups. Of the 12 groups, 5 were in Liberty State Park because of their work; for Liberty Park workers and Liberty Park officials and volunteers, the park is their workplace. Other individuals who were in the park for reasons related to their work were the organized group leaders. The other 49 participants were in the park for their own pleasure. Of the participants, 48 are men and 28 are women; 52 were born in the United States or its possessions, while 16 are immigrants.

Of those consulted, 27 were working in the park. We spoke with six teachers who felt strongly about Ellis Island—one said, "Before the renovation it was like the walls were talking to you. It's beautiful now, but it was more realistic and quieter then. You walked around in reverence and sadness." These teachers had heard of the bridge; one thought construction had begun already. They thought it was "an excellent idea. It's environmentally sound, it will make it more accessible, and draw a larger crowd because of the lower cost. . . . The ferry is cost-prohibitive for some families." One said that approaching Ellis Island on a footbridge would be "a different type of anticipation, like taking a walk back into history. To me, the trip is Ellis Island, not the boat." They thought they could more easily explore the island with their students. They typically are allowed just one trip per year, and the children can only absorb so much in two hours. The other female teacher, a young Jersey City resident, said she "thought they were working on opening the bridge up so that you could walk over it." She said, "I'm a walker—I love to walk. I think it would be great."

A crew of maintenance workers were interviewed together while they were having lunch in the Camp Liberty area of the park. They all opposed the bridge because they feared it would create more litter that would increase their workload. The tunnel alternative appealed to these men as something "cool"—that is, new and interesting. However, the head of this work crew favored the proposed bridge for the opposite reason: he thought the opening of the bridge would bring in National Park Service workers to care for that part of Liberty State Park, thereby reducing his crew's workload.

Three African American maintenance workers felt that the government should be spending its money on families and education, rather than on this bridge. What difference does it make, she asked, if you pay for the ferry directly or for the bridge through higher taxes? One felt that Liberty State Park is "for whites," and so is the proposed bridge, and her coworkers agreed.

Park administrators, on the other hand, were concerned about the increased traffic and parking. One park ranger feared the proposed bridge would bring vandalism to Ellis Island: "I think it stinks," he said. "Who's responsible for people getting hurt?" He also was concerned about wheelchair users and stormy weather, and he felt the Circle Line would not continue its operations from Liberty State Park. He said his main concern is crowds and the kinds of people who would go if it was totally free. Yet a volunteer at the Interpretive Center felt a bridge would serve the poorer people of Jersey City: "They should think of the many people who don't get the chance to see the islands because of the fares. You know, I never saw the Statue of Liberty. My mother never saw it. I didn't see it until I came to work here. I couldn't afford to take my family. You're talking about people here in the city who are in tight circumstances."

Vendors, however, also responded with both positive and negative assessments of the bridge proposal. An African American T-shirt vendor felt the bridge would threaten the vendors, the Circle Line, and anyone who made their livelihood from the present arrangement. "It's not a good idea. ... Tourists get a thrill out of taking the boat, seeing the skyline. ... A bridge will hurt our business ... but it would definitely put the Circle Line out of business." An African American fruit-juice vendor, though, thought it "an excellent idea. More people would have access to the island." She could imagine the bridge "being nice and cool in the summer. ... More exercise, and you can take as long as you want walking across." The third vendor said, "It's kind of wrong that you have to be charged to go over. It's lots of money to some people—they could walk over. There should be another way, even if [the ferry] was subsidized."

Of the 49 men and women who were interviewed while recreating in Liberty State Park, most were in favor of the proposed bridge. Three Spanish-speaking men sitting on benches in the picnic grove felt that a bridge would provide easier access and thought the bridge would afford good exercise opportunities, for jogging, bike riding, and walking. He said the bridge would afford "more choices, something more to do." To another man, the proposed bridge and its alternatives "all sound good—just get the project going; there are future generations and growing populations that need means of traveling." An older African American couple likened Ellis Island to the monuments and institutions in Washington, D.C., saying that Ellis Island should be on the same level, with free access, and "opened up to all to visit. Having to take a ferry is cumbersome, and we all know and have in mind the costs." He thought people would like the exercise and would appreciate the freedom of walking.

A few visitors however, did not want a bridge. An African American man walking with his young son along Freedom Way on a weekday morning felt that the bridge portended an undesirable change of use of Ellis Island from a tourist site to something more intensive, perhaps commercial. Another white man felt that the bridge would compromise the character of Ellis Island. "I think one of the things to consider is the fact that it was an island. [A bridge] would afford people the opportunity to go who wouldn't take the ferry . . . but it's not the same feeling—I definitely think it would take away some of the mystique of the island."

All the fishermen supported the proposed bridge, and two brought up the idea of fishing from the bridge. One man said the bridge could be like Liberty Walk: "It could be beautiful." He said the bridge would make Ellis Island attractive for a Sunday outing with his wife and kids. He noted a significant perceived safety factor in the bridge's favor: "You know something, a lot of people don't want to take the boat. My wife was afraid of it—she thinks the boat could sink, and sharks could get her. Scared." He then referred to the movie *Jaws*. The other Hispanic fisherman, 26 years old, also imagined walking over to Ellis Island with his wife, to explore the buildings and see the view. The fourth fisherman, who was there with his wife and three grandchildren, identified himself as a Filipino and a senior citizen. He said he was disabled. He thought the bridge was a good idea, but felt that a security guard was needed. "If it's open all the time, the wrong people will get in."

For the most part, the constituency groups are not predictive of attitudes toward the bridge—with the one notable exception of vested interest among Liberty State Park officials and workers, who are overwhelmingly against the bridge. Among the other constituencies, the active recreation groups, like walkers and cyclists, would appear to be more solidly in favor of the proposed bridge than the passive user group and the organized group leaders. The passive users may be, as a group, less mobile than the active and thus less interested in being able to include Ellis Island among their activities.

A sharp distinction in cultural orientation emerges from the data between the Hispanic and other interviewees. The Hispanic interviewees were solidly for the bridge and highly receptive to the alternatives; they were generally for development and attracted to the novelty of an elevated train or a tunnel. Although the Hispanic maintenance workers were all against the bridge for fear that it would increase their workload, they also said that they liked the tunnel alternative because it would be new and exciting. Non-Hispanic interviewees, however, were much more skeptical of the alternatives. All the opposition to the bridge came from among non-Hispanic participants, and the non-Hispanic bridge supporters rarely shared the view of the bridge as progress. Rather, if they supported the bridge, in many cases they did so because they would like recreational access to Ellis Island or because they identified with the national cultural symbolism of Ellis Island.

Overall, the arguments for and against the bridge show some differences. Many of the favorable attitudes involve personal preferences, such as "I'm in favor of it," "It would be nice to walk over," "It should be free," "I'd be more likely to go," "Good exercise," "It would be quicker," and "I'd ride my bike." By contrast, excluding the personal views of the eight maintenance workers concerned about their workload, negative attitudes were often based on issues of public policy and cultural values.

Other than Liberty State Park workers, few people opposed the bridge for reasons having to do with the protection of Liberty State Park. We had expected that people who hold particular attachments for the parts of Liberty State Park most likely to be affected by the bridge would have had concerns about the bridge. For example, we thought that the attachments that fishermen or walkers have to Liberty Walk might cause them to be concerned about an Ellis Island bridge. However, all the fishermen and all but one of the walkers whom we consulted were enthusiastic about the bridge. The negative attitudes expressed by non-park-employee participants were more focused on perceived impacts on Ellis Island. However, all four of the Liberty State Park officials consulted had negative attitudes toward the bridge, stemming in part from concerns that ancillary parking needs would constrain the development of future recreation space within the park. A far more pervasive concern among both the recreators and the workers was the impact of the proposed golf course on the park: many participants volunteered opinions on the golf course, hardly any of them favorable.

VALUE ORIENTATIONS FINDINGS

The value orientations of Liberty State Park workers and users were similar to those at Battery Park in terms of the variety of ways that people think about the proposed access alternatives, but their priorities were quite different. Table 4.4 presents our findings listed in order of the number of times that the value orientation was expressed. As at Battery Park, many participants responded with more than one orientation so that the total number of responses (90) is greater than the number of participants who responded (66).

The two most frequently cited value orientations among Liberty State Park users were "health and recreation" (11) and "park quality" (11). Health and recreation includes the responses of those for whom the appeal of the proposed

Table 4.4. Value Orientations at Liberty State Park

Value Orientation	Examples	Number of Responses
Health and Recreation	"I would probably go every day for exercise" "I love to walk; it would be great"	11
Park Quality	"they need a guard to keep the wrong people out"	11
Access	"it will make it more accessible … walking costs nothing"	8
Aesthetic	"there is no view from a tunnel" "it could be beautiful"	8
Cost	"as long as there is no toll" "I think it should be free"	7
Economic	"it would definitely put the Circle Line out of business"	7
Social Priorities	"it is wrong to be charged [a fee]"	7
Choice	"everyone would have options" "they should have another way even if the ferry is subsidized"	7
Political	"now they are tricking it up"	5
Safety and Comfort	"it would be more convenient" "my wife thinks she will drown"	5
New Technology	"the [elevated] train would be something that you do not see every day"	5
Education	"you will be taking a walk back into history"	3
Ecological	"the monorail would have the best impact on the environment"	3
Personal	"for the number of times that I am going to visit the ferry is fine"	3

bridge is in being able to walk or bicycle across to Ellis Island as part of an exercise routine or recreational experience. The health and recreation responses were all elicited from persons who were in favor of the bridge. Park quality, on the other hand, includes the responses of people who are concerned about potential adverse environmental impacts from the bridge on either Ellis Island or Liberty State Park. Nearly all the people who were concerned with park quality were not in favor of the bridge or the other access alternatives.

"Aesthetic" (8) refers to either the appeal or disturbance of views of a bridge, or from a bridge or elevated rail, as well as the absence of views from a tunnel. For instance, one fisherman, looking out at Ellis Island and the existing bridge as he spoke, said he felt that a bridge between Ellis Island and the CRRNJ Terminal area would block views of the water from Liberty Walk. A schoolteacher was concerned that the bridge itself could be an eyesore. Improved access to Ellis Island (8) was a value orientation of those who saw a bridge as providing easier access to the park; however, in one case a respondent was concerned that improved access might increase vandalism.

The next four value orientations—the cost of ferry fares (7), economic impact issues (7), social priorities (7), and increasing the choice of access modes (7)—were cited by people both for and against the bridge. Most of the people who supported the idea of a bridge hoped that it would be free of charge. Participants who mentioned social priorities or economic concerns usually were concerned about the adverse impact of the proposed access alternatives, while people who mentioned their preference for greater choice wanted to retain the ferry and add the bridge, tunnel, or elevated rail.

Political value orientations (5) were expressed by participants in favor of the bridge who suspected that politics will prevent it from being built. Participants who mentioned new technology (5) were in favor of progress; they liked the bridge and in some cases liked the tunnel or elevated rail alternatives even better. People expressing comfort and safety concerns (5) were generally for the bridge since they viewed the boat as a risky proposition. Those who cited education (3) were all in favor of the bridge, while those citing ecological (3) or personal (3) value orientations were divided in their attitudes toward the access alternatives.

Neighborhoods in Jersey City
PHYSICAL AND SOCIAL SETTING

Three neighborhoods bordering Liberty State Park were selected for study: 1) Paulus Hook, a small gentrified area of brownstone row houses and corner parks; 2) Van Vorst, a larger area of elegant brick and brownstone row houses focused on Van Vorst Park, a residential square, with some gentrification amid a highly heterogeneous population; and 3) Lafayette, a mixed industrial and low-income residential area of tenements, wooden row houses, public housing projects, and newer, subsidized modular housing. These neighborhoods were selected both for their proximity to Liberty State Park and because they are representative of the social diversity of Jersey City.

Paulus Hook is a historic, peninsular neighborhood across the Tidewater Basin from Liberty State Park. The center of the neighborhood is made up of three corner parks across from one another, where people frequently sit on benches in the shade during the hot summer afternoons. The park users are representative of the various residents of the neighborhood: some are Polish-speaking immigrants who are longtime residents of the area, some are Spanish-speaking recent immigrants, and a few are older English-speaking European Americans.

The gentrified center of the community is Washington Street, a mixed residential and commercial street, with an expensive Italian restaurant across from law and real estate offices. There are a number of churches in Paulus Hook, including Eastern Orthodox, Roman Catholic, and a Polish Roman Catholic church, Our Lady of Czestochowa. Each of these churches offers numerous community activities and services, including senior centers, parochial schools, and summer children's programs. Paulus Hook has a well-organized neighborhood association that meets on the first Thursday of each month, although not during the summer months. However, members of the association were interviewed by telephone and were included in the residents' database.

Van Vorst, named for the park at its center, includes York, Mercer, Montgomery, Monmouth, Varick, and Barrow streets, all lined with substantial row houses of brick and brownstone dating from the mid- and late 1800s. The largest and most splendid houses look out on Van Vorst Park from Jersey Avenue. Gentrification has been under way in Van Vorst since at least the mid-1970s. Many houses in the neighborhood have been refurbished and their architectural details restored. On the same streets, salsa music can be overheard from double-parked cars of residents who have stopped to talk to a friend at the local bodega or to someone sitting on a row house stoop. Many of these conversations are in a mixture of English and Spanish. Farther down the street, elderly African American residents sit or stand on their stoops conversing with neighbors who are returning home or passing by on the way to the busy bodega. Van Vorst has a number of churches, including various Spanish-speaking congregations of local evangelical groups.

At the center of the Van Vorst neighborhood is Van Vorst Park, a grassy, Victorian square, lined on its four sides with stately townhouses. In the center of the park is a bandstand where concerts are given on summer evenings. Near the bandstand are swings and other play structures surrounded by benches where parents and caretakers sit watching their children.

Lafayette, located along the western edge of Liberty State Park, is a residential neighborhood with many intrusions of car repair shops, scrap metal yards, and piles of old tires and other industrial clutter. Part of the neighborhood has small manufacturing shops side by side with residential streets. Lafayette is bounded on the west by Garfield Avenue, on the north by Grand Street, and it extends to Caven Point Avenue on the south. It is separated from Liberty State Park on the east by high railroad embankments and the elevated New Jersey Turnpike Extension.

Most of the community members interviewed were African Americans or Spanish-speaking Caribbean Americans who had lived in the neighborhood for some time. According to the 1990 census, the median household income is $8,422, well below the 1990 median household income of $29,054 in Jersey City at large. Families live in brick or stone row houses, in larger apartment projects, or in the new subsidized modular attached townhouses. The center of the Lafayette African American male community is the barbershop where men sit, talk, and exchange news throughout the day. The bodegas and bus stops on each corner of Pacific Avenue also provide opportunities for conversations and neighborly interchange, particularly for women, younger men, and mothers with young children. The major school in the area is the Assumption–All Saints parochial school, headed by Sister Maeve McDermott. According to Sister Maeve, she is responsible for 750 children in this relatively poor area. The Convent of the Sisters of Charity has been a mainstay in the community for over 80 years and has run the school and summer programs for local children. There are a number of other churches throughout the neighborhood, including the Monumental Baptist Church, where we interviewed a number of the congregation, and the African Methodist Episcopal Church, as well as several small evangelical and storefront congregations.

Lafayette also includes a two-block enclave of small row houses where older working-class white Americans reside. The Groovy Pub acts as their community center; men spend the day talking both inside the pub and on the corner. Two other local eating places include the Pacific Tavern, frequented by local workers, and The Little Place next to the Catholic church. The community also encompasses Lafayette Park, a pleasantly green oasis of about 25 acres, between Maple and Lafayette streets, across from Monumental Baptist Church. The park has a bandstand, swings and other play equipment, a wading fountain, and tennis courts for community use.

Constituency Group Findings
PAULUS HOOK

Seven people were interviewed in Paulus Hook, three women and four men ranging in age from 10 to 65. Of those interviewed, two were African Americans, one was a Spanish-speaking Caribbean American, and four were European Americans. The residents interviewed had very mixed opinions about the Ellis Island bridge proposal; three said that they supported the plan, while four were opposed.

Two participants prefer the ferry: "I prefer the ferry. I can walk any time. Which is going to cost more? The $6.00 (in 2004, $7.50) for the ferry or the money to build the bridge?" "I like the ferry because I can see the island better." A third person said that she is outspokenly against the bridge. She feels that the people of Paulus Hook "haven't read the fine print, so they will probably say 'yeah, that sounds like a good idea,' without realizing the problems." Another participant was concerned about the increase in traffic and parking problems, in contrast to the residents of Lafayette, who view the increased traffic and use of the park and neighborhood as a positive impact.

Of the three people who support the bridge possibility, one thought that it was a good idea because it would "connect the island and the mainland." Another said "a bridge would be perfect. I'd go to Ellis Island. They are trying to build up Ellis Island . . . where all the people came through. Circle Line is a beautiful line, but it is too expensive today." The third interviewee said: "My *bubba* [Yiddish for grandmother] came through Ellis Island in 1910. You know what that means, don't you? There are lots of lessons to be gained . . . on where we came from, who we are, and so on. It is important to recognize that there is no longer a need for an island location."

VAN VORST

The majority of the 33 people who were consulted in Van Vorst favored the bridge. Those who were positive about the bridge alternative offered a wide variety of reasons for their enthusiasm with comments that were in many ways similar to the themes discussed in Paulus Hook: "It will provide democratic access," "I want a physical link that people can walk across," and "Ellis Island is an important place to visit because of the history." Residents of Van Vorst also observed that it would be great for biking and walking, particularly by people in their neighborhood. One older resident suggested that "young people could walk across it, but I would not use it . . . the ferry is for out-of-towners. The bridge would be better for those in town . . . but it really is not going to change that much . . . that is a dream they [other residents] will have to wake up to." Another said, "A bridge gives you a sense you could navigate so easily." The majority of residents think it is a "great idea." The people that our community consultant spoke to in the neighborhood association meeting also thought that the bridge would be an improvement, but under one condition: "The members of the neighborhood association are for the bridge, but only if it is free. If there is a charge then the bridge will not be accessible to many in the neighborhood."

Those residents who were opposed to the bridge proposal often had complex reasons and concerns. One young woman was concerned that "with a bridge you do not have the control you have now. I was there recently; it was pretty crowded. There would be abuse [of the island] and graffiti." A middle-aged man also was concerned: "I think a bridge makes it more accessible, and I worry about vandalism." Another younger man said: "It is an interesting question. A bridge could exceed the carrying capacity of the island. The ferry creates a natural control over crowding." And one resident felt that the money for the bridge should be used "to fight drugs and build a community center in the area," while another simply said that she felt that it was a "waste of money." In Van Vorst, then, although there is considerable support for the proposed bridge, some residents have concerns about the impact of a bridge on the physical environment of Ellis Island, and some feel that the money could be better spent on other community development and social service projects.

LAFAYETTE

Of the 73 people interviewed in Lafayette, almost all had positive attitudes about building a bridge to Ellis Island, because they could walk there and would be able to visit without paying the $6.00 for the ferry ride. One resident said, "I resent the guy at the end of the bridge that is there so I cannot walk across. The reason I want a bridge is so that the children can walk free." Others commented: "You could walk over and not pay." "It would help poor people take their children." "It is expensive to go with three children." Other residents felt that a bridge would improve the park: "It would help the community because more people would go because it is easier." "More people would come to the park and that would be good." Three residents emphasized that the bridge would offer a nice view. Another added that it would be easier to go to Ellis Island with a bridge, and that more people going was good because it was "an important place to see a lot of things to remind you of how the past was."

A small minority of residents were opposed to the building of a pedestrian bridge. Most of those opposed like or prefer the ferry ride: "I feel more secure on the ferry. The ferry with a child is the best. They can stay with you and not run out of control." "I would miss the ferry ride." One resident was concerned about the bridge itself: "The bridge would be too long." These residents seemed to feel that the ferry would disappear if there was a bridge, while the users of Battery Park and Liberty State Park felt that a bridge would have no impact on the success of the ferry. And seven residents had no opinion about the bridge one way or another; they felt that either they were uninformed or that their opinion did not matter. We tried to explain to residents that their opinion did matter, but not everyone could be convinced.

Overall, residents wanted a bridge so that they could go to Ellis Island. Most had never been there because of the high cost of the ferry and the large size of their families. They would like to go as a family group (or group of friends) to learn and enjoy the view. As Sister Maeve McDermott of the Sisters of Charity put it, she has 750 children whom she is responsible for and who would like to be able to visit Ellis Island for educational purposes. Unless the bridge is free, however, it will not help residents to visit the island. Residents were not

Table 4.5. Neighborhood Value Orientations

Value Orientation	Lafayette	Van Vorst	Paulus Hook	Total
Cost	17	15	3	35
Park Quality	8	9	3	20
Access	9	3	1	13
Health and Recreation	8	0	1	9
Education	3	4	1	8
Community Quality	5	2	0	7
Aesthetic	4	1	1	6
Economic	6	0	0	6
Choice	2	3	0	5
Ecological	0	4	0	4
Political	3	0	0	3
Safety and Comfort	2	1	0	3
Social Priorities	0	2	0	2
Personal	0	1	0	1
New Technology	0	0	0	0

concerned about the possibility of increased traffic in their neighborhood. In fact, many see the increased traffic as a positive good, bringing more people into their community. Residents of Lafayette are very proud of Liberty State Park, and many see the park as the best thing that has happened to the neighborhood in many years. It was clear from the way that people spoke about the park that it is very important in their daily lives, and they would like to expand their educational and recreational horizons to include Ellis Island. As it now stands, Ellis Island is an expensive tourist site that is only visited as a onetime experience with a school group. The ferry ride is viewed as a tourist experience, and not something for local residents. For a middle-class family, a visit to the "islands" is an activity that might take place with out-of-town visitors. But for the majority of residents in Lafayette, most of whom live a ten-minute walk from Liberty State Park, the cost of a visit to Ellis Island is prohibitive.

Value Orientations Findings

The value orientations of the communities are reflected in the differences in priority given to health and recreation (Lafayette vs. Van Vorst and Paulus Hook), ecological concerns (Van Vorst vs. Lafayette and Paulus Hook), and economic concerns and community quality (Lafayette vs. Van Vorst and Paulus Hook), as can be seen in Table 4.5. The residents of Lafayette, in fact, discussed concerns related to access, health and recreation, economics, and community

THE ELLIS ISLAND BRIDGE PROPOSAL

quality more than the other two communities. Lafayette is the closest to Liberty State Park and has the largest number of families living below the poverty line. Thus, their discussions of the proposed changes to the park reflect their need for recreational space and improved community facilities and their very real concerns about local employment. Interestingly, cost and park quality were the greatest concerns for all three communities. When the three communities are treated as one neighborhood, cost, park quality, and access emerge as the dominant value orientations of residents, in contrast to the economic and health and recreation concerns of Battery Park and Liberty State Park users, respectively.

Conclusions

The following conclusions were drawn from the comparative analysis of the three study areas: Battery Park, Liberty State Park, and the neighborhoods surrounding Liberty State Park. Our most important observation, however, is reflected in all the findings: that is, that the people we talked to were overwhelmingly interested in the questions we asked and were quite sophisticated in their understanding of the problem and its consequences, regardless of cultural or educational background. Thus, assumptions that the general public would not be able to evaluate the access alternatives or would not care about the proposed changes to Ellis Island and Liberty State Park were unfounded. This finding suggests that the environmental assessment and planning processes can be enhanced by consulting local populations through the REAP process.

Table 4.6 presents the value orientations compared across the parks and neighborhoods. What is clear from this comparison is that each area has slightly different priorities and concerns. Battery Park workers and users are not at all concerned with the cost of the ferry or the bridge, but instead are concerned about the possible economic consequences of the proposed access alternatives. Liberty State Park workers and users, on the other hand, are concerned with the health and recreation advantages and park quality disadvantages of the access alternatives. The residents of Lafayette, Van Vorst, and Paulus Hook are most concerned with the cost of the ferry or proposed access alternatives. Cost, access, park quality, and economics were the most frequently mentioned concerns for all groups. Table 4.6 is useful in understanding the variation among these populations and can be referred to as a way to judge how often a concern was expressed by participants in this study.

The bridge is overwhelmingly the preferred access alternative for reasons of safety, cost, ease of access, choice of time and space, and health and recreation benefits. Because of safety and cost concerns, almost no participants thought that a tunnel is a good idea. A few participants thought that an elevated rail

Table 4.6. Value Orientations: Comparison across Parks and Neighborhoods

Value Orientation	Battery Park	Liberty State Park	Surrounding Neighborhoods	Total
Cost	0	7	35	42
Access	13	8	20	41
Park Quality	6	11	20	37
Economic	23	7	6	36
Health and Recreation	9	11	9	29
Choice	9	7	5	21
Aesthetic	6	8	6	20
Social Priorities	10	7	2	19
Political	8	5	3	16
Education	4	3	8	15
Personal	8	3	1	12
Safety and Comfort	4	5	3	12
New Technology	5	5	0	10
Ecological	2	3	4	9
No Impact	9	0	0	9
Community Quality	0	0	7	7

might be fun or exciting; however, they added that it would be too costly and might break down. Participants from all economic groups were negative about the proposed subsidized ferry because of issues of cost, crowding, and governmental intervention.

A large proportion of participants, especially those from low-income areas and those who were interested in the needs of low-income families, were concerned about cost issues, including the high price of the ferry and a possible charge for the proposed bridge.

The differences in attitudes toward the proposed access alternatives were not predictable by constituency group. Instead, there was a marked difference between the attitudes of immigrants and native-born participants, and between the attitudes of people who work rather than recreate in the parks. The native-born participants and the workers were more concerned about the negative impact of the proposed access alternatives. Native-born participants were skeptical about the political decision-making process and the social priorities reflected in the decision to build a bridge. Workers were concerned about losing their jobs or profits, or the negative impact of a bridge on the quality of the park.

People perceive the potential impact of building a bridge, elevated rail, or tunnel on the Ellis Island experience in similar ways, but they interpret that impact very differently. For instance, all groups agree that a bridge would increase the number of people who would visit Ellis Island, but those who are opposed to the bridge see this change as a negative impact because of crowding, while those who are for a bridge see this change as a positive impact as more people will learn about history.

The same is true for the perception of what will happen in Liberty State Park. Most participants agree that the proposed bridge or other access alternative will increase traffic and the number of people in the park and in the surrounding communities. Some of the participants, such as those in Paulus Hook, view these changes as negative, while others, particularly those in Lafayette and seniors, welcome these changes. Increased traffic means inconvenience for some and increased economic potential for others.

Residents see Ellis Island as a place of recreation as well as of history. When a bridge is added it is perceived as providing a nice place to walk, with wonderful views, as well as access to learn more about history. They feel that the ferry and Ellis Island are primarily for tourists and first-time visitors whereas a bridge would be for the local populations who do not normally visit the island. If a bridge is built, then local populations would visit Ellis Island more frequently. Teachers, parents, and neighbors would like to learn more about Ellis Island and the history of the region and nation.

The importance of this study cannot be underestimated. Allowing poorer residents the ability to visit could change their sense of stewardship and entitlement. And there are access solutions that do not require the building of an expensive bridge. If the issue is that these residents be able to visit in groups and families, and the ferry is too expensive, then subsidizing a Sunday ferry for local residents would provide a solution to what is seen as a highly contested change in the landscape. Further, the recreational aspect of having a bridge could be addressed in many other ways. Understanding cultural values opens up many avenues and solutions to local problems in ways that can resolve even the toughest of problems facing park planners and administrators.

Chapter 5

Jacob Riis Park
Conflicts in the Use of a Historical Landscape

Introduction

If small parks and plazas in the city are the primary focus of William H. Whyte's questions about social viability, then this book expands the scope of spaces to be considered by examining large open spaces, including urban beaches. The importance of urban beaches to questions of social viability has been obscured by concerns for traditional urban spaces such as plazas and "green" areas like the neighborhood park and community garden. Yet, privatization and commercialization processes have had an impact on urban beaches. New commercial environments—especially shopping malls—attract people away from beaches and other traditional leisure-time resorts. These seductive, air-conditioned spaces promote recreational consumption but, through surveillance and prohibitions, sharply limit the variety of public encounters and self-expression.

In contrast, public urban beaches are places where more socially and culturally diverse populations encounter one another while engaging in a great variety of activities. In federal parks the government is required by law to consult the public and to consider the viability of cultural groups' lifeways when managing and making major changes in a park. Seashore parks play important roles in the continuity of cultural practices of a number of urban communities; they also support social sustainability and fortify democratic processes. Yet, little has been written about the cultural ecology of urban beaches. When beaches are discussed in scholarly works, the focus is on the natural ecology of beaches, tourism, real estate value, and development. Little is known about how beaches function as social places and as parts of the urban landscape.

In this and the following chapter, we examine two public beaches. Following Whyte, we ask why one appears to be a more viable social space than the other. Although the social life at Jacob Riis Park is lively, visitation declined sharply in the 1990s and remains lower than park managers would like it to be. This chapter examines the issues at play as park management grapples with an underutilized facility in transition—in particular, the conflict between the recreational preferences of current visitors and the historic preservation mandate.

Jacob Riis Park is a beach with a boardwalk, playgrounds, food concessions, and a surprising history. Its location on the Rockaway barrier-island peninsula, in the borough of Queens, is closer to Brooklyn than it is to most of Queens. Riis Park began as a municipal park named for the prominent social reformer, Jacob Riis, who became famous for his photographic documentation of life among immigrant children in the Lower East Side slums of New York. An advocate of parks and recreational facilities for the urban poor, Riis was influential in the construction of Columbus Park, one of New York's first Progressive-era playgrounds, on the cleared site of the notorious Mulberry Bend slum in Lower Manhattan. Accordingly, Jacob Riis was honored by the construction of this park on a Long Island barrier beach that complements ocean swimming with athletic and recreational facilities.

In 1974 the National Park Service inherited Jacob Riis Park from the New York City Parks and Recreation Department and incorporated it within Gateway National Recreation Area. While the Park Service has made some improvements at Riis Park, it has lacked sufficient funds to reverse the long-term decline of the park's developed facilities. According to Billy G. Garrett (2004, personal communication), when Jacob Riis Park was deeded to the federal government, it was in terrible shape. Something on the order of $15–17 million was spent by the NPS on stabilization and rehabilitation of the primary facilities, and estimates for the remaining work are in the vicinity of $10–12 million. The lack of subsequent funding, he argues, is due to the fact that construction funding typically comes out of a ranked list of agency needs, so the issue is not whether Congress will fund the work, but whether the level of funding is sufficient to allow inclusion of the Jacob Riis projects. We, on the other hand, attribute this lack of funds to an insufficiently strong political constituency in Congress for what remains essentially a local beach.

Most users of Jacob Riis Park today are recent immigrants who do not necessarily speak English or have the citizenship or standing to demand facilities or services. Ironically, the cultural behavior of these park users poses a challenge for park staff: their unintended destruction of natural resources through intensive use of the park grounds threatens the integrity of the historic landscape and adds to the challenge of restoring the park on a limited budget. This situation raises the question of how park administrators can balance the competing needs of visitors and historic preservation.

In the summer of 2000 the Public Space Research Group was asked to conduct a rapid ethnographic assessment procedures study at Riis Park for the National Park Service as part of an effort to understand the decline in park usage. The Park Service was undertaking some physical improvements to facilities and grounds and considering changes in the park's concessions. Park managers wanted to know more about their new immigrant users and how to meet their social, recreational, and cultural needs.

When we first visited the park, we were surprised at the park's deteriorated condition. Considered a good example of art deco, interwar-period public recreational architecture, much of Jacob Riis Park is listed on the National Register of Historic Places. Improvements and modifications to buildings and grounds are therefore subject to the exacting requirements of historic preservation practice. I (Dana Taplin) wondered whether the National Park Service was unduly burdened by historic preservation requirements in maintaining the park. In our initial trips to the site, we conducted key-informant interviews with park personnel. One June morning Suzanne Scheld, Larissa Honey, and I traveled to the headquarters of the New York section of Gateway—located in a drab, former military office building at Floyd Bennett Field—to interview Deputy Superintendent Billy Garrett. Early on in our conversation, I asked Mr. Garrett whether he thought that the National Park Service had been burdened with spurious National Register designations that diverted scarce funds needed to keep its facilities in good working order. I asked, "Is this something that Park Service wants? Did the impetus to landmark these things come from within the Park Service or is that something that you feel sort of stuck with and have to work around?"

An architect with a good measure of historic preservation work in his Park Service résumé, Mr. Garrett's first response was a chuckle. He then said, "No, I don't. . . . I would not suggest that I'm stuck with them; I wouldn't characterize it that way." After a pause he defended the Riis Park listing in the National Register, in which he stressed the importance of documentation and professional evaluation in creating landmarks:

> I take very seriously our natural resource base and our cultural resource base. What I am interested in is insuring that, when someone says to me "This is a resource of either cultural or natural significance," that they've got the data to back that up, and that they have gone through the process for evaluation. Now, the properties that we're talking about right now as cultural resources—and specifically Riis—there's not any question at all in my mind that that process has been gone through and that as a manager we can work forward and make decisions that incorporate that set of values.

Fair enough—but the historic designation constrains management flexibility in adjusting to new uses and cultural values among users, and maintaining these historic structures is expensive. But the historic designation also has positive implications in the areas of funding competition, public relations, and branding. At least within Riis Park, the historic significance of the buildings and grounds relates to the park's mission of public recreation. Elsewhere in Gateway, the Park Service is at pains to find uses and maintenance funds for abandoned military facilities and other "cultural resources" that relate indirectly if at all to the idea of a national recreation area.

Gateway National Recreation Area is an assemblage of formerly city-operated public spaces like Jacob Riis Park, other beach and tidal wetland areas, an ecologically important wildlife refuge, and several large, surplus military facilities. Finding new uses for abandoned military facilities is always a challenge. At Gateway and at other locations around the nation, such facilities have been transferred from the Department of Defense to the Department of the Interior, to be managed by the National Park Service. Once designated as park lands, the disused facilities can be rehabilitated, coupled with adaptive reuse that is sympathetic with important characteristics of the building. The reuse formula, however, usually calls for preservation and conservation of these facilities as cultural resources, to complement the Park Service's natural resources, and similarly involves visitor centers, interpretive programs, and other forms of curatorial management.

Even with the National Register of Historic Places listings, the Park Service has still struggled to find suitable uses and adequate maintenance funds for its substantial collection of "cultural resources" at Gateway. Floyd Bennett Field is a good example: here is an early municipal airport associated with the famed aviators Amelia Earhart, "Wrong Way" Corrigan, and others that later served as a naval air station during World War II. Surely there are good reasons to designate this place as a historic landmark, but at the same time, visiting an old airfield at the far reaches of Brooklyn is not high on most people's lists of things to do in New York. How can the Park Service find the funds to turn the old terminal into an attractive visitor center? What can be done with the cavernous hangars? What should happen to the concrete runways? The Park Service has sought answers to these questions since the inception of Gateway in 1974. For example, according to Billy G. Garrett (2004, personal communication), they include information about the field and its history in the educational and interpretative programs. They also find uses for the field and its holdings that are compatible with the historical significance of the field, its associated natural resources, and the mission of the park. On that basis, visitors enjoy activities

Figure 5.1. Jacob Riis Park bathhouse, promenade, and beach

ranging from gardening and flying radio-controlled model airplanes to cycling, bird-watching, and camping. At the same time, we found that some of the hangars have been used for years by the helicopter unit of the New York Police Department—a use that makes sense except for the premise that this is supposed to be a park.[1]

At Riis Park, the boardwalk, the bathhouse, and other buildings along the boardwalk, the parking lot, and the landscaped grounds constitute the Jacob Riis Park Historic District—essentially the whole park. Billy Garrett told us "there may be some non-historic intrusions, so to speak, that would not be considered part of the historic district"—playgrounds, for example, added since the "period of significance" (the 1930s)—but they are few. The effort to tether the park to a "period of significance" leaves management less able to be flexible in adapting to changing needs. While a nonhistoric playground can be rebuilt or eliminated, the historic 72-acre parking lot, which is more than half empty on even the hottest summer days, must be preserved. As of 2000 the Park Service had spent $15 million to reconstruct portions of the bathhouse structure in keeping with preservation standards, but long-term development plans and lack of funding do not allow for indoor showers and changing rooms.

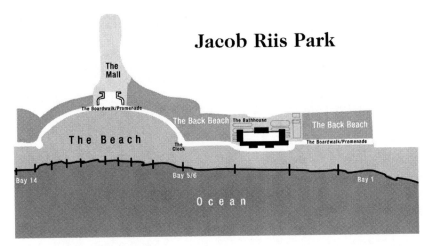

Map 5.1. Jacob Riis Park

Picnicking—and people and cultural groups who take part in this activity—is the point of sharpest conflict at Riis Park between contemporary park use and a management policy based on historic preservation. This chapter focuses on how a REAP study helped to understand the new immigrant users so that the park could begin to find ways to accommodate their picnicking needs while at the same time protecting the historic landscape and providing the necessary services for long-term beach users.

Methodology

The scope of research required that we provide an overview of cultural groups using the park, including an analysis of their concerns and the identification of cultural and natural resources used by and/or culturally meaningful to them. The information would assist managers in evaluating alternatives and requests for access, as well as in assessing impacts of proposed changes in park design and programs on local users.

Rapid Ethnographic Assessment Procedures

A number of rapid ethnographic assessment procedures methods were used during the various phases of the research process. Individual interviews were completed in Spanish, Russian, or English, depending on the preference of the interviewee. The interviewers had a map of the park available for noting any specific site and to stimulate discussion about Jacob Riis Park. Our goal was to involve and educate as well as to interview community members about the park design process.

We collected 131 interviews, divided almost equally among six sites. The sites included Beach Bay 1, Beach Bays 5–6, and Beach Bay 14 and three designated picnic areas: "the Mall," "the Clock," and "Wood Park," each named for the closest landmark. The sites were selected to represent the different cultural, ethnic, and activity-based groups that use the park. Behavioral maps of each site sampled both weekends and weekdays from 8 a.m. until 8 p.m.

Expert interviews were collected from park staff and volunteers identified as having special expertise to comment on current park problems. These interviews were completed during the first week of the project to help identify park concerns and problem areas. Some expert interviews were followed up with transect walks.

Table 5.1 reviews the methods used, how much time was spent on each, the kind of information that was produced, and what was learned.

Table 5.1. Jacob Riis Park:
Methods, Data, Duration, Products, and What Can Be Learned

Method	Data	Duration	Product	What Can Be Learned
Behavioral Mapping	Time/space maps of site, field notes	5 days	Description of daily activities on-site	Identifies cultural activities on-site
Transect Walks	Transcribed interviews and consultant's map of site, field notes	3 days	Description of site from community member's point of view	Community-centered understanding of the site; local meaning; identification of sacred places
Individual Interviews	Interview sheets, field notes	20 days	Description of responses of the cultural groups	Community responses and interest in the park
Expert Interviews	In-depth interview transcriptions	5 days	Description of responses of local institutions and community leaders	Community leaders' interest in park planning process

Historical and Social Context

The idea of creating a noncommercial public beach on Rockaway beach followed a period of beach resort development that began with the opening of a railroad across Jamaica Bay to Rockaway in 1880. Rockaway became part of New York City with the municipal consolidation of 1898. The city began seeking land for a public beach in 1904 and purchased much of the present site in 1912. The site was diverted for use as a naval air station during World War I. Riis Park's largest structure, the bathhouse, was opened in 1932, shortly after the navy withdrew across Rockaway Inlet to Floyd Bennett Field. Although it lacks the swimming pools of the Jones Beach prototypes, the Riis bathhouse was modeled on them. The new park also had handball and tennis courts, playgrounds, and landscaped grounds (Wrenn 1975; Lane, Frenchman, and Associates 1992).

This early history notwithstanding, the Jacob Riis Park of today is a product of the career of Robert Moses, the great urban planner who changed the face of mid-twentieth century New York. As parks commissioner beginning in 1933, Moses took an early interest in Riis Park. His office's plan for reconstruction, issued in 1936, transformed the site. Like many Moses projects, the reconstruction plan was linked to major road projects—in this case, the new Marine Parkway Bridge connecting the park with Flatbush Avenue in Brooklyn and another Moses project, the Belt Parkway. Visitors coming by car over the new bridge could park in a gigantic new parking lot. In fact, the whole plan was a tribute to automotive movement: the swift, triumphant approach over the new bridge above Rockaway Inlet, then a short glide along the backside of the beach on a looping, six-lane parkway that leads to the parking lot entrance, itself set on an angle across the curving parkway.

Moses built concession buildings and a new, elliptical boardwalk. The concessions buildings articulate the start of a grassy pedestrian mall visually aligned with the Empire State Building, which is visible on the horizon. The landscaping and pedestrian circulation system was completely redone. Japanese black pine trees, the most distinctive element of the Riis Park planting plan, were a personal favorite of Moses. Disparaging the design of the then four-year-old bathhouse, Moses had some of its Moorish architectural ornaments removed and the towers raised for dramatic effect. A beachfront pavilion that projected too far toward the water for his liking was torn down and replaced with a streamlined, modern facade, and new extensions were built at either end. Moses masked the resulting stylistic mixture by painting the building white, which he ordered for all the structures at Riis Park to unify the aesthetic effect (Wrenn 1975).

JACOB RIIS PARK

Under the supervision of the New York City Parks and Recreation Department, Riis Park thrived, attracting visitors from local Brooklyn and Queens neighborhoods as well as from Manhattan. William Kornblum, a sociologist working with a team of researchers for the NPS, documented the cultural life of the beach and boardwalk in 1975. Kornblum and his colleagues describe a lively boardwalk with members of the culturally diverse neighborhoods strolling by, interacting with one another, and enjoying the ambience and everyday life of an urban beach scene. He documented how each of the many beach bays that are divided by jetties created distinct territories of different social groups, including a gay beach at Bay 1, an African American crowd with parties and music at Bays 5 and 6, and local Italian Americans from Bay Ridge occupying Bay 14. Kornblum and his colleagues discussed how this socially complex environment worked by providing separate "territories" for the different cultural groups, while at the same time allowing for social intermixing, girl-watching, break dancing, gambling, and card playing by all groups on the boardwalk that runs behind the beach area. Even today, people talk about the vitality of the boardwalk and marvel at how well it worked in conjunction with the bay beach social structure to reduce intergroup fighting and social conflicts and encourage a friendly and fun atmosphere for apartment-dwelling New Yorkers to visit in the summer. Today Bay 1 continues as a gay beach, but the composition of Bays 5 and 6 reflects the demographic changes over the intervening quarter century in Brooklyn and Queens, with increasing numbers of new immigrants settling there from all over the world. This change has disrupted some aspects of "Black" place attachment cited in the Kornblum report. On the other hand, these new populations are rapidly adopting Riis Park, symbolically "marking" it and making it their own.

The New York City Department of Parks and Recreation managed Riis Park until 1974, when it became part of the new Gateway National Recreation Area. While Gateway has achieved some of its ambitions as an urban nature and recreational resource, Jacob Riis Park has not fared well within the National Park System. Jacob Riis Park was in a condition of severe deterioration when it was transferred. The park's handsome bathhouse has been out of service since 1978, and many of the original buildings and food kiosks are closed. The Park Service has restored portions of the ship railing along the boardwalk, but elsewhere the railing is badly decayed. The wooden walking surface was removed years ago, leaving only a coarse concrete "boardwalk" to walk on.[2]

Most athletic facilities in the "back beach" area, such as ball fields and paddleball courts, are so deteriorated as to be unusable, and one of the children's playgrounds made a newspaper list of "worst playgrounds in New York" the year of our research. Today the historic bathhouse is being renovated again, and there are plans to restore the Mall and other elements of the Moses design. The findings of our study, however, suggest that restoring the landscape may not accommodate the people who currently use the park—some of them users described by Kornblum and others more recently arrived in New York. We describe the areas of the park studied and our findings concerning users' values and activities in each of these areas. As we talked to people in different sections of the site, conflicts become quite apparent, whether it is a need for more bathroom or changing facilities, more barbecue grills and picnic tables, or a cleaner, more family-oriented environment.

We interviewed users in two distinct kinds of environments: the bays, beaches, and jetties studied by William Kornblum and the "back beach" areas where picnicking and playground equipment is located. We focused on three beach bay areas—Bay 1, Bays 5–6, and Bay 14—in order to study the changes since 1975. In the back-beach areas we were concerned with sampling the major picnicking and play areas favored by the new Latino users, including Wood Park, the Mall, and the Clock areas (see map 5.1).

Settings and Findings: The Bays, Beach, and Jetties

Jacob Riis Park beach is unusually wide because the dunes were removed at some point in the area's history as a public beach—probably during Robert Moses's reconstruction of Riis Park in 1935–1937. The beach is narrowest at Bay 4, opposite the bathhouse, although not very narrow even there.

The jetties that separate the bays at Riis consist of a wall of reinforced concrete that is mostly buried in the sand, and then a double row of wooden pilings that extend 40 feet or so into the water beyond the low tide line. At the jetties, large boulders have been piled up around the point where the concrete ends and the wood begins. The Riis Park jetties do not attract people to walk or sit on them, and the lifeguards direct swimmers away from the jetties when they get too close. Signs at jetties where the pilings stand lower in the water warn people away from "submerged objects."

Like other barrier beaches on Long Island, the sand at Riis Park is very fine. It seems browner than at beaches farther east—perhaps some kind of construction sand has been mixed in during beach replenishment. In many places the sand is mixed with very short, reedy sticks, as well as many shell shards. The beach seems quite clean of trash for an urban beach. The use of heavy mechanical equipment for daily cleaning leaves the surface very flat, with long, straight tracks and grooves left in the sand.

Bay 1

The trash that collects along the back side of the beach and this bay's relative isolation certainly don't stop people from coming to Bay 1. On a sunny weekend day this bay is filled with hundreds of people. Even on an overcast weekend well over 200 people are crowded into Bay 1, although at nearby Bays 3 and 4, which are more centrally located, very few people are present. The crowd is quite diverse, although the makeup changes with the day of the week and the weather. On a sunny weekday there are more women on the beach and more groups with both men and women. Many people bring their own umbrellas, and most are clustered in groups of two or more, mostly more, which tend to be segregated by sex. The groups are well supplied: they have coolers, many have chairs, and some have radios or boom boxes. The music is not as loud here, however, as it is at Bays 5 and 6. On a sunny day several of the women are topless; however, we did not see full nudity at the beach, as Jacob Riis is not a clothing-optional beach.[3] Far fewer women are present on the overcast weekend, and only men are observed on the rainy weekend.

In general, this bay is an adult area. We only saw children or teenagers on the sandy area once, though a few older kids sometimes play in the water. Most of the people were sunbathing, socializing, and eating. Many of the people knew one another at this bay—there's a lot of interaction between people from different groups here, and many know each other by sight. The bay is ethnically diverse, although there is a predominance of blacks and Hispanics. The bay also attracts mostly adults in their 20s, 30s, and 40s with a smattering of older and younger people.

The 18 males and six females interviewed reflect the large number of male visitors. Most were in their mid-20s to early 40s, creating a predominantly young adult space with few children, and identified themselves as black, white, or Hispanic. Three-fourths of those interviewed were born in the United States and all were well educated.

Bay 1 visitors overwhelmingly say that this area is meaningful to them as a space of freedom and liberation. Several say that this beach is very significant because it is one of the few beaches with a gay area where they feel comfortable and safe to express themselves. Many people also associate this beach with good times spent with family and friends. Several say they have met many good friends here over the years, many of whom they now consider family. The beach is also meaningful as a place to escape the city and enjoy a day of leisure with a sense that "you're living larger, the good life." Yet, there also is a feeling that this bay has been abandoned or neglected by the park management, as indicated by the absence of a lifeguard, bathrooms, and concessions and by the abundance of trash and debris on the beach.

There is a sense of territoriality and place attachment associated with this bay. "This is a gay and lesbian beach—it's our beach" is a sentiment heard over and over. Several people say they love this bay and that they would be very sad if it were closed down. This territoriality is also expressed in terms of outsiders entering the bay. Occasionally an outsider will come in to gawk or cause problems, and several people who come here regularly make it their business to keep an eye on the visitors and to talk with newcomers in order to make sure no conflicts occur. In this sense, territoriality is not expressed in terms of exclusivity.

Bays 5 and 6

In Bays 5–6 people sit singly or in groups, spacing themselves at more or less consistent intervals from other parties. The most popular part of the beach for sitting and sunbathing is the first 20 to 30 feet above the high-tide line. The closest spacing is parallel to the water, because most visitors want to be close to the water, and perhaps because people tend to orient themselves toward the water rather than facing up or down the beach. Thus on a busy day, visitors will sit farther away from the people behind or in front of them than from the people beside them. It's rather like shorefront property, where the lots tend to be narrow and deep.

Many people bring folding chairs to sit on and coolers containing food and drinks. Some sit directly on a blanket or towel, or lie down to sunbathe. Many parties bring an umbrella for shade, a few put up something more substantial, like a tent.

Bays 5–6 are relaxed and quiet, and loud boom boxes are not much in evidence. The crowd is mixed—white, black, Hispanic; families, young singles, older folks.

The boardwalk at Bays 5–6 has two sets of outdoor showers, the Clock, and the food concession and restrooms at the west end of the bathhouse. It provides access to the playground, ball courts, the official barbecue area, and a path to the bus stop and parking lot. There are flat benches with no backs on the seaward side of the boardwalk and regular park benches with backs on the landward side. The Riis boardwalk was described in Kornblum's 1975 report as a kind of flamboyant happening, somewhat like the famous Venice Beach in Los Angeles. Now it is much quieter. Still, people stroll along the boardwalk and sit or hang out in groups on its benches. The surface is actually concrete, the boards having been removed 30 or more years ago. One consequence is that the

Figure 5.2. Picnickers at Jacob Riis Park

Figure 5.3. The Clock at Jacob Riis Park

boardwalk is a popular shortcut for official and unofficial vehicles. The pedestrian is frequently disconcerted by the sound of some oversized van roaring up behind him—sometimes a succession of such vehicles. The concrete surface is more suitable for bikes than an actual boardwalk, and lots of cyclists come and go, including bicycle-mounted Park Police officers.

We collected 23 interviews, which reflected the even ratio of men to women. The visitors were predominantly Brooklyn residents, high school rather than college graduates, middle- and working-class, and in professional and semiprofessional, municipal government, and union trades occupations.

Bays 5–6 comprise the most popular, crowded section of the Riis Park beach. There are many regulars here, partly because the Clock serves as an important orienting landmark. A number of people say they meet their friends here or that their friends will know to find them here. Historically, Bays 5–6 were identifiably African American. It is much less so now, but remains popular with African Americans and Caribbeans. There is an extended group of African-heritage families and individuals who hold a proprietary attitude toward Bay 5. These people have been coming here for many years. One woman there said, "We have a big attachment to Bay 5. Blacks started coming here to Bay 5 in the 1930s. My grandmother came, my daughter. We have a tremendous love for this space: It's ours." One member of this group said they number about 150 on any given Sunday. Some of the longest-standing, most-devoted members hold informal titles, such as the "King of the Beach." We interviewed the "Prince of the Beach" and, later, another man who declared himself the "Mayor of the Beach." The Mayor has been coming here since the 1940s, and he travels now with his family from their home in the Maryland suburbs of Washington. Many members of this informal club are well known to the rangers: they greet one another like old friends and chat on the beach or along the boardwalk. The members we spoke with are unhappy with the deterioration of the park and the decline in attendance. Some remember livelier times when Tito Puente would come here and perform. One of them remembers a plan to put in a performance facility here, and a Caribbean-born woman thought live music here would be great—as long as it didn't attract the wrong crowd. Some members of this club have met with park officials to express their views. One man, coming from a meeting with Deputy Superintendent Garrett that morning, said, "We're back to square one. Nothing's maintained or refurbished. No hoops; there's grass growing up in the basketball courts. The bathhouse is closed. I have not seen any improvements in two years. The money is appropriated; where is it going?" To the Sandy Hook (New Jersey) section of Gateway NRA, he thought, where "they have seven beaches—beautiful, new everything." Another member noted that it's a wealthier community out there. These men spoke of misplaced priorities; they thought the park should renovate disused existing facilities like the bathhouse before putting in a swimming pool.

On a brighter note, several visitors to Bays 5–6 commented on the availability of the back-beach areas—that Riis is a beach with a park attached to it. That means that there is room to cook out here comfortably, and there are playgrounds for the children. African Americans, in particular, like the availability of ball courts at Riis Park, although they note that these facilities are in poor condition. A few people thought the boardwalk surface should be smoother and kept free of sand.

People here most often say they come to Riis Park because of the convenience—it's not far from their homes in Brooklyn or Queens—and the parking is reasonably priced and always plentiful. Visitors like the mellow atmosphere: people get along well; they don't fight. One regular visitor said, "There's a lot of love here." She also found it very safe, in part because of the alertness and skill of the lifeguards, and she was not the only one to express appreciation for the lifeguards.

Bay 14

At Bay 14 the boardwalk has no facilities and acts as a sort of cul-de-sac for boardwalk activity as well as official surveillance. As such, it is a redoubt of groups of beer-drinking young adults who hang out here undisturbed for hours. On any Sunday there were groups of people, sometimes in pairs, and others in groups of five or six members, mostly men. Some have been regular visitors over a period of many years. They stand around on the concrete surface near their coolers, drinking and socializing. The crowd in Bay 14 is nearly all white, although not exclusively so, but mixed in age.

The 20 interviews we collected in Bay 14 had half again as many men as women, and were predominantly with white people from Brooklyn. Bay 14 had more high school than college graduates, and professionals and semiprofessionals, city workers, and union trades people were the dominant occupations among persons interviewed. Bay 14 has a higher proportion of frequent visitors, people who come several times a week, than Bays 5–6 or Bay 1. Sunbathing and swimming were the main activities.

As in Bays 5–6, visitors in Bay 14 most often say they come because of the relative convenience of Riis Park to the Rockaways or nearby sections of Brooklyn. For several people, however, the appeal of Bay 14 goes beyond convenience. A number of visitors had very strong attachments to this bay, left over from its seventies heyday as a teenage hangout. A retired policeman from Bay Ridge, in his early 40s, had grown up coming here and said he still knew people here, "like the guys standing over there on the concrete (boardwalk): neighbors, friends, people I grew up with and met through coming here. This side of the beach is all Brooklyn; down there, Queens. They used to call this the Bay Ridge bay." His wife, also a retired police officer, said it "used to be every weekend, this was the place. We used to hitchhike to the beach from Bay Ridge! Yeah, all the time, a group of girls. Put out your thumb, got a ride. Right on the Belt Parkway." Another person said that Bay 14 used to have a big teenage scene in the seventies. A fireman in his 30s said he has been hanging out on the boardwalk here with his friends for 18 years—"in this exact spot!"

Most visitors thought the beach should be kept cleaner and that trash baskets should be put back on the beach. Some of these comments were more specifically addressed to the beach, which looks much less picked-up than other areas. One woman asked why the rangers weren't distributing trash bags the way they were last year. Yet others thought it was pretty clean. A firefighter thought the lack of trash baskets left the beach dirtier than it should be but it was still "pretty pristine." He also thought the Park Service had done a tremendous job of sand replenishment here in June. A Russian woman liked the "nice sand, clean sand." The retired policewoman thought the water here is very clean, "even during the needle summer!" (1988). She added, "It's so important—the parks and the water—it really is, to have a nice place where everybody can go."

People like Bay 14 because it is not as crowded as other parts of the beach. Their biggest complaint is of the lack of services: no showers, no food, and no proper restroom. A woman complained that the women's restroom at the Mall was filthy, and a few people commented on the lack of lifeguards at this end: one thought that led to people bunching up in the bays that do have lifeguards. In some ways, visitors here like the low official presence in Bay 14: the groups of regulars believe that alcohol consumption is legal in federal parks and liked the fact that they can drink beer on the boardwalk here undisturbed (actually, alcohol is illegal in city and federal parks alike).

Settings: The Back-Beach Areas
Wood Park

"Wood Park," our informal name for a playground with wooden play equipment, is located east of the bathhouse, adjacent to the East Lot parking area for handicapped visitors, the boardwalk in front of Bays 3 and 4, a small open field on its eastern side, and the park drive that exits toward Neponsit. In the very center are three tall wooden play structures set in a sand box enclosed by a low, thin cement curb. On the drive side, there are several groups of backless benches close to the playground, and beyond these lies a cement pathway lined with trees. On the other side of the play structures lies an open space and some pieces of weed-covered asphalt that suggest there was once a facility situated here. An isolated water fountain is also placed in this undefined area. Closer to the parking lot is a low chain-link fence that separates two sets of swings from the center of the playground. The swing seats and ground padding have been removed, yet the metal frame remains standing.[4] No trees are located in the center play space of Wood Park, and the ground is a mix of sand and bits of grass. At the margins of the playground, near the parking lot, is a narrow island of healthy grass. Nearby lie other grassy enclaves somewhat enclosed by trees and bushes; these areas were not formally planned for shading picnickers, although this is the manner in which they are generally used (figure 5.2).[5]

On summer weekdays this area of the park is very quiet. There may be a handful of picnicking families, usually from Brooklyn or Queens, whose members happened to have a day off from work and decided to spend it in the park. They come with barbecue and beach equipment and generally sit under the trees closest to the boardwalk. Women may also be found in the play area supervising small children on the play equipment. On weekdays these visitors appear to park in the East Lot, ordinarily a restricted parking area.

On a sunny, hot weekend day this area is transformed into an active family barbecue area. Families generally arrive around 8:00 in the morning and stay in the park until the early evening. Throughout the morning caravans of cars, small vans and 4x4s can be seen pulling up to the curb in front of Wood Park. The passengers of these vehicles jump out and begin making several trips back and forth to the grass as they unload their bundles of grills, coolers, lawn chairs, grocery bags, baby carriages, party balloons, mini sound systems, and an occasional visitor in a wheelchair.

The perimeters of the playground and the spaces closest to the parking lot are where most of the trees are located at Wood Park; therefore, the first groups to the park generally set up for the day in these areas. Some visitors who want to sunbathe, or who come to Wood Park with a tent, choose to settle in the open field. Others looking for space and shade not far from the parking lot settle in the grassy areas across the drive from the playground. Adults guard the spot they have staked out, while teens and children tend to roam around the park. Throughout the day a black 4x4 with government plates slowly cruises through the area on the cement paths. From the car window, park rangers and EMTs advise visitors about some of the park rules and answer the questions of visitors who approach their vehicle.

In general, weekend picnickers are groups of families and friends that number from 6 to 15 people per group. Often, many of these groups cook out in conjunction with two to five groups of other families; thus, the total number of some parties can range between 30 and 75 individuals. Many of the groups include persons representing two or three generations of a family.

The Clock

The Clock area is located adjacent to the boardwalk at Bays 5–6, west of the bathhouse, opposite Riis Park's old-fashioned street clock. The area lies between a baseball diamond and a playground that has sprinklers.[6] The picnicking area has seven cookout grills in the space closest to the beach. Wire trash bins are scattered throughout the area, and a large red metal drum for coal disposal is situated near some of the grills. The Park Service delivered new picnic tables to the area during the summer. Now the area has approximately a dozen new wooden tables in addition to two fairly worn-out ones.

The area has a combination of shaded and open spaces. The area closest to the boardwalk is predominantly without shade although there are a few small, twisted black pines and several dead hardwood trees planted several years ago that did not survive in the sandy soil. The back wall of a closed down concession booth on the boardwalk and the walls of a closed-off concrete tool shed

Figure 5.4. Park visitor cooking in shade cast by concrete wall, Riis Park

near the ball field fence also provide a small amount of shade. The space closest to the parking lot with many tall trees has lots of shade, a floor of pine needles, and big branches from which to hang hammocks, party balloons, and family reunion banners. Beyond this area, in the direction of the bathhouse, is some open lawn space and an enclave of low bushes that cast a bit of shade.

The Mall

The Mall area includes the long, rectangular formal lawn flanked by walkways, nautical lampposts, and benches near the boardwalk. It also encompasses shaded clearings behind the buildings of Bay 9, and open islands of grass and bushes in front of the handball courts. The lawn space and shaded areas on the east and west sides of the Mall are fenced off for restoration. Visitors, however, have knocked down the green wire fencing in several places in order to get access to the shade trees.

This area of the park is heavily populated on the weekends and sparsely visited on the weekdays. On the weekends, families and groups of friends arrive as early as 8:00 in the morning with the intention of staying until late in the afternoon. These families are predominantly Hispanic, Russian, and Caribbean of African and Indian descent. Picnickers come with lawn chairs, blankets, coolers, and portable grills. They set up underneath the available shade, including the thin areas up against the new fences, and under the few inches of shade along the walkways. Along the edges of the Mall lawn the shade is limited and thus many groups spread out from the fence onto the formal pathways. Some tie string hammocks between the lampposts and trees located within the fenced area currently off-limits to visitors. Others clear out spaces deep within the bushes behind the brick building that stands between this area and the boardwalk. After people leave, piles of dumped charcoal, patches of burnt grass, and paper plates remain. Some visitors who prefer the sun set out blankets and chairs on the open lawn, while others come prepared for the lack of shade and set up puff tents and portable cocktail tables with umbrellas. In the area of the Mall closest to the parking lot, volleyball games take place around a net set up by the visitors.

Findings: Back-Beach Areas

A total of 64 interviews were collected: 18 in Wood Park, 24 in the Clock barbecue area, and 22 at the Mall lawns, the back beach, and the boardwalk. An equal number of males and females were interviewed between the ages of 21 and 50. In terms of findings, half of the visitors consulted identified themselves as Hispanic. Nearly a quarter identified themselves as black American, and nearly another quarter as white European. Half of the visitors consulted are new immigrants from Central and South America, eastern Europe, the Caribbean, or the Middle East. The other half were born in the United States or Puerto Rico. The majority of the Spanish-speaking immigrants were born in El Salvador, Guatemala, or Colombia. The white Europeans were born in Russia or Poland and the vast majority of the U.S.-born visitors identified themselves as black American or Hispanic. The majority of visitors interviewed live close to the park.

About half of the visitors reported that they had completed high school or grade school, or had never attended or completed any level of education. A small portion of these individuals are youths who are still attending school. A third of the visitors reported that they had obtained college degrees. The visitors reported that they worked at a variety of professional and service-sector occupations such as teaching, nursing, livery service, factory work, and domestic labor. Economically, the visitors are a mix of middle-class, lower-than-middle-class, and poor families. Of the individuals who reported their income and number of household members, a significant number live in large households that are supported by limited incomes. One Puerto Rican woman identified her income as "low, low, low, the lowest you can have."

A third of the visitors interviewed have been visiting the park for more than 10 years. Many said that they grew up coming to the park. Another third reported that the day of the interview was their first visit to the park, and some had only started to visit Riis on a regular basis over the last two summers. Most of these back-beach picnickers come in large groups. More than half come with their family and friends, and nearly a third come with informally organized groups of families. Sometimes these groups of families visit the park to celebrate a family reunion, birthday party, or an informal get-together of coworkers. Others had come to the park with a formal organization, including a church group and a men's shelter.

Their primary activities include picnicking, socializing, relaxing, supervising children at the playground and on the beach, attending parties, and visiting the beach for a walk or to people-watch. They also swim, or play basketball, paddleball, or baseball. Some mention playing sports such as soccer and volleyball for which there are no designated or permanent facilities.

Visitors like the park because they have a sense of attachment to the place. They also like the beach and the convenience of the park. Many found the park to be clean, well maintained, and a good place for children to play. They like that the facilities are closely spaced and therefore easily accessible, enabling them to supervise young children and creating a sense of safety in the park. Many visitors also enjoy the atmosphere of the park and its beauty. They say they feel "comfortable," "relaxed," "content," and "free of worries" in the park.

Barbecuing close to the beach is the main attraction. Many park users claim that barbecuing is not possible at nearby public beaches. Some visitors even suspect that, although they were themselves barbecuing, it was not actually allowed. During the interview, some visitors revealed that they feared the volunteer ranger was coming over to put a stop to their party, or they asked the interviewer if what they were doing was in accordance with the rules.

Having said that visitors enjoy the barbecuing opportunity, many visitors are frustrated that there aren't more grills for cookouts. One Hispanic immigrant commented that just because a lot of poor people come to this park the park shouldn't be poorly equipped. Many cried, "We need grills!" At the beginning of the data collection period, visitors also complained about the lack of picnic tables and benches. The installation of new wooden picnic tables in mid-July, however, prompted many visitors to comment on how the park staff was working to improve the conditions of the park.

Visitors also complained about the lack of available shaded space throughout the back-beach areas, and expressed hope that the managers "don't take down any more of the trees!" They asked, "Why doesn't the park develop a program to replant trees every five years?" Shade is a primary amenity for barbecuing, and visitors feel that shaded spaces are too scarce and the areas too limited.

Long-term visitors to the Mall area were particularly upset about the roping off of shaded spaces that have been used for picnicking over the past several years. Some recognized that trees were being damaged, and they usually blamed the carelessness of other visitors for the deterioration of the trees. Nonetheless, they were not happy that the management had closed them out of "their" spaces. One family who was interviewed as they were leaving the park claimed that they would not return as a result. Several families who typically picnic in the Mall were interviewed in Wood Park and the Clock areas, and they talked about being unhappy about having to shift around and use another area.

Shaded areas aren't the only limited good in the park; spaces close to bathrooms, beach, and parking as well as secluded spaces also are in demand. An African American mother explained that she needed space to get away from others because her children get into other people's stuff, and an Ecuadoran father didn't want to bother others with his barbecue smoke. Several Russian visitors remarked that they prefer to be alone and hoped that others would not bother them. Hispanic families and a Hispanic church group claimed to need more space in order to create cohesion among the small subgroups of their large parties.

The limited shaded spaces in the park force many to arrive in the park at early morning hours just to reserve a spot. One young man, for example, was dropped off at 7:00 a.m. with his sleeping bag in addition to a card table and cases of soda in order to hold a spot for his family. The early morning competition for park space is a reflection of how valuable park spaces are to the visitors. The visitors who obtained the "prime spots" commented positively on the fact that there were spaces in the park that enabled them to be close to the facilities and underneath shade. Many visitors, even if they didn't obtain shade, still liked the fact that the close spacing of picnicking facilities, the beach, bathrooms, and playgrounds was an attractive feature. The propinquity and diversity of spaces in the park accommodate children and adults at the same time. "Grass for the adults and sand for the kids," is how one woman put it. "My wife and I can relax here while the kids play on the beach," said a Puerto Rican father swinging in a hammock while drinking a beer.

The majority of visitors in the back beach come to the beach with children. "It's all about children!" a mother explained. "I want to make my children happy, give them a place to play, get them out of the apartment . . . get them some fresh air." Mothers like the park but feel there is a great need to develop equipment areas, activities, and programs that will involve children of all ages. In Wood Park families bring equipment to entertain their young, including a portable sprinkler that one family hooked up in the playground. Families on the Mall said that the playground areas are too far away, and they hoped for more activities for youth nearby. They had come to the Mall only because the other parts of the park that are closer to the playgrounds lacked shade. Mothers also appreciate that the park is safe so their children can roam without a problem. One mother thought that some cars that drive too fast around the park could be dangerous.

Most visitors who came by car did not find the parking lot services too expensive, but many thought it unreasonable to charge visitors more than once a day, as people sometimes like to leave for a while and then return. Some visitors were frustrated by the abuse of parking privileges; they suspected that individuals who weren't qualified made use of the closest parking lot, meant for those with handicaps. One longtime user whose spouse is handicapped said that the convenient handicapped parking lot was one of his main reasons for coming to the park, which was his favorite place. Handicapped parking and drop-off points are important. Visitors with a lot of recreational equipment unload their cars at the back-beach curb in front of the picnic areas. Several visitors said that they understood that parking in these areas is not allowed, yet they expressed a need for assistance in accessing the park. They dislike this park rule that restricts loading and unloading and also feel that the park staff takes an unnecessarily hostile approach to enforcing the rule.

Overall, the back beach is heavily populated by new immigrants whose first language may be Spanish, Russian, Polish, or Hindi, and who may have limited English-language skills. These visitors may be unfamiliar with or have difficulties understanding park rules. They may also be accustomed to using public park spaces in ways that are appropriate within the context of their homelands and cultures, but in conflict with National Park Service rules and the protection of park resources. For example, many Hispanic visitors enjoy relaxing in hammocks; however, this practice may be harmful to the trees, lampposts, and fencing to which the hammocks are tied. Many Caribbean visitors enjoy listening to music at high volumes, and this practice both attracts and distracts other visitors. Many eastern European visitors prefer minimal interaction with park staff and other visitors, a preference that sometimes creates conflict for park staff whose job it is to relay important information about park use to these visitors. Beachside picnickers from countries with warm climates and hot beaches are used to enjoying natural shade and the protection of constructed shelters. In the absence of shading, such visitors often string tarps between tree trunks, throw mats of cardboard up into the tree branches, or prop up tents all over the Mall area. These practices are in conflict with rules to protect the park's natural resources as well as the security of the visitors.

But at the same time, many visitors in the back-beach areas have little to no education and come from large households, with low incomes; in general, they have little time to enjoy the beach. These characteristics, while not true of all visitors, do suggest that many park users live in economically difficult situations. The park offers one of the only forms of recreation locally available and is an extremely important source of release and escape from the stressful conditions of the urban environment and everyday life.

Conclusions

Overall, Jacob Riis Park serves an amazingly diverse population of users, from poor recent immigrants who are coming for the first time to wealthy professionals who have visited the park for more than 20 years. Household income, education, and occupation is evenly distributed, so that from a socioeconomic point of view, it is truly a park for everyone. Even the distribution of first-time and once-a-summer users (32 people out of 131, or 24%) and frequent users—people who come one to four times each week (46 people out of 131, or 35%)—is somewhat balanced. There are more long-term users than first-time users (25 people have come for more than 20 years compared to 17 first-time users), but the park still attracts newcomers while it retains long-term users from local neighborhoods. Jacob Riis is a regional park that serves the populations of Brooklyn and Queens, and clearly it draws upon the cultural and economic diversity of those communities.

At the same time, different users claim distinct territories, and these territories and users have distinct needs and desires. In order to develop a plan of action for renovation and change, it is important to understand that the different territories of the park—the bays and back-beach areas—have clearly articulated, yet distinct concerns. From the perspective of planning and design, it is difficult to develop one proposal that meets the needs of all constituents. Instead, a series of strategies and directives are necessary, each distinct in its objectives and aimed at a discrete population.

For instance, the back-beach picnickers are the newest visitors to the park. These visitors—many of whom speak only Spanish and have recently come to the United States from Central and South America—are the poorest and least able to provide services and resources for themselves. They need more picnic areas equipped with tables, shade (trees, tents, or cabanas), and grills. Bathrooms for the large number of children and elderly in these families, safe playgrounds so that children can play nearby and thus under supervision, and adjacent beach lifeguards are all necessary for their visit to be optimally successful and satisfying. These newcomers prefer to visit as large groups of families and friends. They prefer shade and are accustomed to hanging fabric for shade and hammocks for seating. Currently the historic landscape does not accommodate these visitors' desire for shade and large gatherings. The limited number of dying trees must be replaced with an appropriate substitute for the disease-prone black pines that provide little shade. There were very few picnicking areas in the original 1936 plan. Robert Moses intended that visitors would sit on blankets on the beach and never envisioned (and never wanted) large groups of people barbecuing at what he designed as a middle-class park. In order to accommodate the new visitors' needs, creative designs and better communication between the park staff and these users will be necessary to solve this culture/landscape dilemma.

Further, immigrant visitors are concerned that their children learn as much as possible about their new life in the United States, so programs including swimming and safety are desirable. But many of the parents cannot read or write English, so letting them know about the programs requires specifically targeted and innovative forms of cross-cultural outreach. In contrast to the typical beach bay user, the back-beach population has no interest in the park concessions: they bring and cook their own food. Finally, these groups enjoy music and dancing—especially Latino rhythms and salsa—and would enjoy summer afternoon concerts that remind them of home (and bring a bit of home to their new beach). Because these newcomers are the poorest visitors, and a population that Jacob Riis can effectively serve, we believe their needs should be the park's highest priority.

Unfortunately as a "national" park Jacob Riis does not have the funds for local community development and outreach. But most national parks are struggling with existing funding levels. In a way, the conditions at Gateway are in some cases better than some national parks, and worse than others. In 2000 the park began a four-year program to improve deteriorated urban recreation facilities. Approximately $450,000 was spent to upgrade ball fields, children's playgrounds, restrooms, and picnic facilities. A portion of these funds was spent at Jacob Riis. At the same time, the New York City Parks and Recreation Department's operating budgets have been reduced during this past decade. Still, it seems important to question what would have happened if Jacob Riis Park had remained a city park or been retained in some form of partnership. It might have had more success at gaining the necessary funds for educational and cultural programs like Orchard Beach, since its users have a voice in New York City politics.

Users of Bays 1, 5–6, and 14, however, would prefer not to see limited park resources distributed in this way. They have no interest in picnic areas and educational or cultural programming. Instead, to them, short-term changes should focus on lifeguards, bathrooms, and trash cans. Because these visitors identify with just their particular favored part of the beach, lifeguards are a top priority. For a variety of reasons—gay identity, tolerance, and safety issues—Bay 1 visitors have relatively less flexibility of movement at the beach. Therefore, when there is no lifeguard these visitors are at risk. While immediate funding for another lifeguard may not be feasible, an arrangement could be made whereby a lifeguard would be posted at Bay 1 several times a week. Perhaps a rotation system could be implemented rather than shortening the guarded area of the beach, or lifeguards could be placed in strategic locations, that is, where there are larger crowds. Based on interviews with visitors and observations of the beach, Bay 1 often is left without a lifeguard at times when much-less-frequented areas have them.

People using the beach—rather than the picnic areas—felt strongest about the condition of facilities behind the beach: the rough concrete on the boardwalk with weeds coming through the cracks and sand drifts on the surface, the deteriorated ball courts, the closed bathhouse. Few beach users were concerned with picnic tables, grills, playgrounds, or music. These visitors come to Jacob Riis for the beach and swimming, so their concerns focus on the cleanliness of the beach, the availability of bathrooms and showers, and in the case of Bay 1, having a lifeguard nearby. The focus is on individual rather than family-oriented activities.

Most visitors to the beach bays complain about the lack of restrooms or showers and about the limited varieties of food at the concession stands. More such amenities would likely attract new users. For example, users like the quiet, secluded atmosphere at Bay 14, but providing the basic amenities—bathroom and outdoor shower—would not compromise its seclusion, and there is more beach to the west for people who want real seclusion. Parking is also a problem at Bay 14: the parking lot is quite far away.

The REAP study at Jacob Riis Park uncovered the conflicts that arise when cultural and social groups compete for very limited resources in a restricted, historical landscape. But at the same time, the study shows that Riis Park has succeeded in attracting a wide variety of users by offering diverse territories, or "niches," that different groups can claim. Even though the different constituencies would choose different improvements, all agree that it is a wonderful park that accommodates their activities and cultural patterns of park use with minimal intergroup conflict. And on the spacious boardwalk, the park's most public space, everyone can come together to experience the park's diversity.

Jacob Riis Park works as a beach that serves recent immigrants and poor to middle-class residents of Brooklyn and Queens. The spatial organization of the beach bays and back-beach areas creates territories that encourage a strong sense of stewardship and place attachment. Moreover, these "territories" promote social tolerance and cultural integration at the level of the site. Expanding upon these existing strengths of the park is one way of attracting new users and increased visitation.

Notes

1. Hangar B, however, is the scene of a volunteer program devoted to restoration of ten historic aircraft, and the four hangars at the north end are being rehabilitated.

2. Real wooden boardwalks are a common seaside amenity in the Middle Atlantic states of New York, New Jersey, and Delaware—something people expect to find at public beaches.

3. Efforts to stop nude bathing at Bay 1, Riis Park, culminated in a court case that went to the highest court in New York State. The Court of Appeals upheld a state law banning nude bathing on public beaches statewide.

4. Wood Park has since been expanded, and the original timber equipment has been replaced.

5. By 2004 the old play equipment had been replaced. The trees that made this a popular picnic area in 2000, Japanese black pines, had all died of disease.

6. In 2004 the baseball field has become a large extension of the Clock picnic area. The Park Service built several sheltering pavilions for picnicking groups and has provided many more picnic tables. There are fewer trees than in 2000, however, as most of the remaining black pines have succumbed and no new trees have replaced them.

Chapter 6
Orchard Beach in Pelham Bay Park
Parks and Symbolic Cultural Expression

Introduction

On the Fourth of July in 1996, I (Suzanne Scheld) made my first visit to Orchard Beach. Typically, this holiday conjures up images of barbecues, festive good moods, and the colors red, white, and blue. That day I found all of this in the park. The colors of the American flag were prominently displayed. More often than stars and stripes, however, I saw triangles, rectangles, and crosses—the red, white, and blue symbols of the Puerto Rican and Dominican flags. These markers of Latino and Caribbean identity were tied to tree branches and to posts in the picnic area. Narrow pieces of twine were drawn between tree trunks so as to create private spaces and to delineate one family's picnicking space from another's. These decorative and expressive forms of territorialism impressed me. The park became a common ground for diverse cultural groups to share and to reshape one national holiday.

Orchard Beach is our second case study of an urban beach. In contrast to Jacob Riis Park,[1] Orchard Beach is a well-utilized park. It is thriving, full of life, activity, and cultural expression. Orchard Beach, located in the Bronx on the border of Westchester County, is part of Pelham Bay Park, New York City's largest public space (map 6.1). While it is visited by many, it is especially enjoyed by Latino visitors, seniors, and naturalists. This chapter describes the symbolic expressions of these cultural groups, and it suggests that the number and types of cultural symbols displayed underscore how deeply attached visitors are to the park. The design, planning, and management of a park can stifle the cultural expressions of visitors. In the case of Orchard Beach, however, a laissez-faire approach to management—which is at times intentional and other times unintentional—enables visitor groups to elaborate unique symbolic displays of their connections to the park.

As a thriving public space, Orchard Beach has additional significance. It is a resource for bolstering Latino community identity and thus contributes to sustaining New York's cultural diversity. In this light, Orchard Beach is similar to American Beach, a Floridian seashore park celebrated for its role in African American history. Located on Amelia Island in the northeastern corner of the

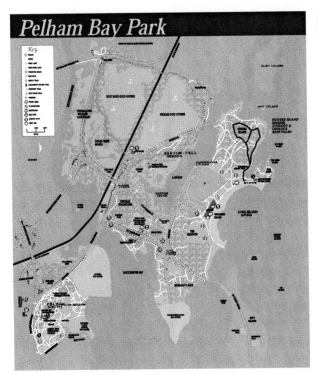

Map 6.1. Pelham Bay Park

state, American Beach was developed as a resort for employees of the Afro-American Life Insurance Company in the mid-1930s, when segregation in the United States prevented blacks and whites from sharing public recreational resources (Phelts 1997; Rymer 1998; Cruikshank and Bouchier 2001). Today residents decorate buildings and murals in the town to demonstrate that the political and social importance of American Beach has not been forgotten.

At Orchard Beach an association with Latino identity was not created in response to the kind of racial segregation that shaped American Beach. However, the economic marginalization and spatial segregation of Latinos living in the Bronx played a role in the development of a strong Latino identification with Orchard Beach and the adoption of the park as a place invested with cultural significance. This chapter reflects how visitor groups, Latinos in particular, communicate cultural meaning through symbolic forms of expression such as music, dance, food, recreational activities, and ways of talking about knowledge and experience of the park. These kinds of communication reflect how marginalized groups within the city can feel "at home" in a park and how they can make it a place of their own.

ORCHARD BEACH IN PELHAM BAY PARK

Figure 6.1. Promenade at Orchard Beach

Methodology

The Public Space Research Group (PSRG) conducted a user study of Pelham Bay Park between July 1996 and June 1998. Orchard Beach was a major focus of the study, although other parts of the park were also examined, including the separate "southern zone" along Interstate 95 and Rodman's Neck.

Periodically park managers conduct user studies to better understand the populations they serve and to efficiently target efforts toward creating a successful public space. Typically these studies focus on describing the demographics of the visitors and collecting user evaluations of the park through widely distributed, self-administered surveys. The park managers of Pelham Bay were open to a complementary cultural approach that entailed extensive participant observation, behavioral mapping, key informant interviews, and face-to-face interviews in addition to a census survey (see Chapter 8 for a fuller description of methods).

The personal connections managers have with the park appeared to be a factor in their receptiveness to an ethnographic study. The administration and staff revealed in key informant interviews that they themselves had strong cultural ties to the park. Many related using the park for exercise and for family get-togethers. Several talked about growing up in the park using particular playgrounds and the beach. Others spoke fondly of the community groups they worked with and of particular park users with whom they came into frequent contact.

During the key informant interviews it became clear that park managers have many stories to tell of people they encountered in the park. Most importantly, even when these stories reveal how park users criticize managers or find fault with the park, park managers related a fundamental appreciation for people's stories. In short, the park has deep meaning in the managers' personal and professional lives, and it appears that their attachments to the park go beyond the scope of their jobs. Managers' social and emotional links to their worksite primed them to find value in an ethnographic study of their workplace.

A second factor contributed to park managers' interest in an ethnographic study. No more than a handful of administrators oversee Pelham Bay and Van Cortlandt—two of the largest parks in New York City—in addition to a host of neighborhood parks in the Bronx. The Bronx park management is undeniably understaffed. They are acutely aware that at times there are management and maintenance issues that they are unable to attend to as a result of limited budgets and staff. The limited contact managers have with the public appeared to take on a special meaning. At the time of the study, managers expressed strong impressions of the park constituents, but they wondered if they gave too much weight to the views of visitors who independently voiced their complaints and suggestions. What about the views of other visitors who, for social and cultural reasons, were not in the practice of voicing their opinions? In defining management priorities, did park management take into consideration a broad range of visitors' views? Park mangers were interested in using the study to verify their personal impressions of the park in addition to deepening their understanding of visitors' views and behaviors.

Given the administration's openness and interest in the park, the study focused on the social ecology of the park and the culturally influenced values and behaviors of park users. The research team conducted fieldwork in Pelham Bay Park throughout the four seasons, during various times and days, and in a variety of weather. At Orchard Beach three targeted areas were selected for close observation: beach sections 1 through 9, the games area (which was partially undergoing restoration at the time), and Hunter and Twin Islands (including the Ecology Center situated on the promenade). Beach sections 10–13, the meadow picnic and lagoon areas, and the parking lot were explored to a lesser extent. The research team examined user's values, perceptions, and preferences through interviews conducted with 149 individuals. Interviews probed users on patterns of visitation, activities engaged in while in the park, knowledge of the park, meanings associated with the park, and evaluation of the park (see appendixes for interview schedule). The patterns of responses were then examined in relationship to gender, age, education, income, race, and nationality.

The user study included a census count, which was conducted at four points of entrance to Pelham Park.[2] At Orchard Beach the counts were conducted in the parking lot. Because the whole park was the focus of our study, the data were not disaggregated to specifically reflect visitation at Orchard Beach with exception of a few particular days. The highest count of visitors to Pelham Bay Park occurred on the Fourth of July with approximately 31,050 persons. An estimated 21,650 users arrived at Orchard Beach that day. The census found that Hispanics/Latinos are the majority of users at Pelham Bay (63%). African Americans account for the next-largest proportion (23%). These findings are in keeping with the demographic characteristics of the Bronx, and they appear to echo visitation at Orchard Beach.

Historical Background and Social Context

Orchard Beach has a history of attracting people who like to make it a place of their own. The park was created in the 1880s, after several years of lobbying by an early environmental group, the New Parks Movement, led by activist John Mullaly (Schnitz and Loeb 1984). Until 1885, when New York City annexed the territory, the entire area east of the Bronx River belonged to Westchester County. At the time, authorities of New York City were opposed to creating a park that might provide more benefit to another county. A series of three mayors were reluctant to sign off on the bill that would create the park. Even after the lands came under the control of the New York City Parks and Recreation Department in 1888, politicians still preferred to use the land for constructing hospitals, prisons, and sanitariums.

The New Parks Movement, however, held firm to their position that the lands would best serve the public as a "pleasure ground." But what sort of a "pleasure ground" would this park be? By the turn of the century, the area was marked by deteriorating farmlands and dilapidated estates, properties that were once owned by the descendants of the sixteenth-century Dutch and English settlers, and before that, the Native Americans (Siwanoy-Lenapi). As the agricultural value of the region declined over the years, the estates and farms were abandoned, one by one. In the eyes of early environmentalists, such an expansive territory gone wild from neglect appeared as lands that had returned to their original, pristine state. The notion of "pristine" wilderness suggested to them a ready-made park, one that would require little intervention. In comparison to other constructed urban parks such as Central Park and Prospect Park, the New Parks Movement argued that Pelham Bay Park required little development and would be less expensive to maintain (Schnitz and Loeb 1984).

According to a history of the park (New York City Parks and Recreation Department 1986), once the city administration took charge of Pelham Bay Park and Orchard Beach, a laissez-faire planning approach went into effect. As little as possible was spent on building facilities and park development in general. The Flying Lady, a monorail built in 1910 to improve the accessibility of Orchard Beach, exemplifies this reserved approach toward park development. Before the days of widespread automobile use and beyond the reach of the streetcar system, the park was difficult to reach. The Flying Lady helped to overcome this obstacle by shuttling visitors through the wooded areas between the Bartow railroad station and the City Island Bridge. After a brief period in operation, a car tipped off the tracks, and while the accident was not serious, the monorail was never restored nor was another means of transportation implemented. Orchard Beach remained difficult to access for those without private means of transportation. Paradoxically, at the turn of the century, the laissez-faire approach to park management and development, an approach that was intended to preserve the beauty of the park for the masses, in fact excluded the majority of people from enjoying it.

Budget constraints contributed to the park's low-key approach toward management during the early years. From the very beginning, the Parks Department relied on "partners" for development. The administration rented out the mansions as hotels and restaurant concessions in order to produce revenue. On the one hand, this age-old "cost sharing" approach allowed for private concerns and volunteer groups to collaborate with the park. On the other hand, interested parties then had the power to shape park resources as they saw fit. Charitable acts could sometimes contradict the image of the park as a public reservation. For example, the administration depended on organizations such as the International Garden Club, which in 1915 fully restored Bartow Mansion and its splendid English gardens (New York City Parks and Recreation 1986).

The park's passive approach to management eventually favored an increase in visitation and visitor autonomy. The completion of the Bronx and Pelham Parkway in 1911 gave the public more access to the park, and soon enough there emerged the well-known "Tent Colony," or "Tent City," as it was also referred to, a community of summer-long campers who set down roots in the park. Originally, the campers lodged on Hunter Island in the early 1890s. In that era the campgrounds were quaint and quiet. According to historical records, in 1907 the parks commissioner issued approximately 250 permits to campers who occupied the thick forest and muddy, rocky shores of Hunter Island. By 1922 the popularity of camping at Hunter Island had expanded: more than 534 permits were issued for more than 3,000 campers (Lubar 1986, 76). Rapidly over time the volume of camping increased to the point that visitors taxed the natural resources, forcing the administration to shift camping off of Hunter Island and over to Rodman's Neck, where it continued to flourish until the 1930s.

Camping in the Tent Colony took on a life of its own. At the end of the summer, their return to next season was assured in that the park administration allowed campers to store their equipment in the public bathhouse until the next summer (ibid.). The structure of the tents they inhabited indicated a certain permanency of these visitors in the park. The camps were nearly solid shelters on wood and cement foundations. Some had wooden walls while others were made out of stretch canvas. On the average, they were as large as 20 feet by 40 feet, wired with electricity and telephones, and decorated with brick walkways, furniture, and Chinese lanterns (Scott 1999, 91). Visitors went so far as to give their tents special names such as "Spare Time" and "Idle Four" (92).

Historians document the community solidarity that was practiced within the Tent Colony. As described by Henry Lubar "[T]he beach had offered a serene and communal atmosphere. Strict code enforcement by the Health, Fire and Parks departments ensured safety, and rules were, for the most part, enforced by the campers. . . . The campers organized their own street-cleaning, fire-fighting, and life-saving corps, and all garbage was buried about a mile away in a specially created fill" (1986, 76). One camper's memoir corroborates this image of community cooperation. He writes about emptying portable commodes at the Tent Colony. Residents would pass each other along Shore Road, either coming or going, carrying their oversized "pots" (Sims 1986). The total lack of comment as each resident passes other residents indicated that the "trip" was an accepted aspect of living in camps at Orchard Beach in 1916 (Sims 1986). The sense of community was also expressed in the private services that campers organized for themselves, such as the daily delivery of a variety of foodstuffs and the mail; and in the organization of community festivities such as lawn parties, croquet tournaments, and games for children (Scott 1999, 91).

The spirited Tent Colony was also known for its strong sense of American nationalism during World War I. Campers hung the American flag on their tents and supported the soldiers who at the time were training in the park. In 1917 the United States Navy set up a Reserve Training Station within Pelham Bay Park on Rodman's Neck, an installation that would mark the beginning of a long history of the park's accommodation of various uniformed services within its borders.[3] At the time, navy sailors brought additional spirit to the park, and campers responded by cheering them on from the banks of the campground. They would send buckets of cold lemonade and jelly water over to the training grounds as gifts for the sailors (Scott 1993, 53).

According to historians, the Tent Colony was an animated, self-organized community strongly rooted in the park. However, the community eventually came to an end. Overcrowding, overuse, and the development of a negative public image propelled the gradual intervention of park management to evict the community in 1935. There were many complaints from the public about the Tent Colony. People complained that permits were only issued to Bronx County Democratic Party workers (Lubar 1986, 78), and there were suspicions that favoritism was at work as many campers were city employees. Others complained that park permits were used to turn profits by subletting sites. Members of the colony confirmed this phenomenon when they complained that an increasing number of "undesirables" were allowed into the community (ibid.). In short, as the Tent Colony expanded, the community itself fragmented and the public increasingly viewed the happenings at Orchard Beach with a critical eye. In order to restore a positive public image to the park, a change was in order for the Tent Colony.

In the 1930s, the park administration swiftly shifted from its laissez-faire approach to management to top-down decision making. It was decided by a new administration that the park was no longer to be a nature preserve that bred anarchy. It was to be developed as a recreation area open to a broader public. In 1934 Robert Moses became the commissioner of the Parks Department in New York City. Moses's first move at Orchard Beach in 1935 was to raze the Tent Colony as well as a handful of other recently constructed structures that he did not approve of. He then began his plans for redesign which included importing tons of fine white sand from the Rockaways and New Jersey in order to create the 1.2-mile-long crescent-shaped beach and to connect Hunter and Twin islands by landfill. He also constructed a new and modern bathhouse that contained locker rooms for 7,000 people, a dance floor, performance space, restaurants, and concessions. In front of the bathhouse he created a formal mall entrance, and between the building and the beach he implanted a 50-foot-wide promenade. Other features included several playgrounds and shuffleboard, basketball, handball, and paddleball courts. The largest of the facilities constructed was a parking lot with space enough for 6,800 automobiles. The facelift was intended to expand and modernize the park, to make it a recreational area for families and a public space for exciting events to take place. Programs such as Wednesday night fireworks, beauty contests, dances, concerts, and calisthenics classes contributed to the upbeat and healthful atmosphere of the park that Moses aimed for (Lubar 1986, 80). Impressed by the new luxurious facility, local users gave it the name "The Riviera of New York City." Today many still call it the Bronx Riviera.

Figure 6.2. Pelham Bay from the Orchard Beach Promenade

Figure 6.3. Concessions area at Orchard Beach

Restructuring the park succeeded in drawing more visitors to it in the post–World War II period. During this time, various groups of predominantly Western European heritage used the park. Seniors today remember how these groups made the park a place of their own. Sections 11–13 of the beach are reported as the former hangout of Italian and Italian American youth, and Hunter Island is remembered as the site of the "Bavarians from Yorkville."

One contemporary visitor of Italian heritage recalls how the "Italian" girls from Arthur Avenue were an exciting but "rough" crowd. Because of her light skin, she "passed" for "a Schneider or Reinhart" and was able to use other sections of the beach. Another visitor remembers that although camping was no longer a formal activity in the park, Bavarians and other visitors inhabited the park in the wintertime. She states:

> [A]s a kid I would come to the park to sleigh ride . . . that's when I'd see them cooking with big pots . . . delicious smelling stew. . . . We'd die from hunger smelling their stew, it smelled so good. They had red cheeks from drinking so much Schnapps and from sitting around a fire. They were friendly people who camped up on Hunter Island all year round, but particularly in the winter. I guess this is what the weather in their homes was like.

Other contemporary visitors remember the American nationalism of some of the visitors who made victory gardens in the park. In an interview conducted near the Ecology Center, one senior reports:

> Years ago people from 86th Street . . . those people from Europe . . . the Germans, Czechs, and Hungarians . . . they'd come up here and camp on the weekend. . . . They'd build volleyball courts in there . . . no one would touch them . . . they made beautiful gardens . . . those gardens weren't vandalized either . . . they' be there all summer . . . you'll never bring that back to this park. . . . Did you see the rocks, the remnants of the courts they built? Did you notice the rock foundations? They're in those alcoves where people like to fish now . . . they were terrific places. . . . Have you been in there?

In the 1960s and 1970s the composition of the park's visitorship permanently changed. In the Bronx the Latino population expanded at a faster rate than that of African Americans and Asian Americans, due to a host of sociological factors, such as increased Puerto Rican migration to New York and the gentrification of certain Manhattan neighborhoods that resulted in a shift of minorities to the Bronx and other boroughs. As neighborhoods in the Bronx increased in density and in minority populations, many socially mobile white families moved to the suburbs. Additionally, the nearby rooted middle-class white communities in Westchester County restricted Glen Island and other beaches to Westchester residents, thereby diminishing the choices of shorefront recreation for Bronx residents. The combination of Bronx migration patterns and middle-class property enclosures contributed to driving up the numbers of visitors, and in particular of Latinos at Orchard Beach.

The 1960s and 1970s were also a time when the United States underwent a period of major social change. Traditional authorities were questioned, and public social behavior was liberalized. Manifestations of these changes were visible in urban parks in the form of experimental drug use, open sex on the beach, and frequent escalated conflict among visitors and between visitors and park staff and police. Budgets were tight at the time and it became difficult for the park administration to get control of the numerous problems, including criminal and violent activities taking place in the less-supervised wooded areas. The park's loss of control over visitors resulted in a temporary negative image of Orchard Beach and its visitors. Such problems were not unique to Orchard Beach, however, as many urban shorefront parks throughout the country were experiencing similar difficulties. For example, similar developments were occurring at Jacob Riis Park in New York (Kornblum 1975) and in Los Angeles (Edgerton 1979).

Since the sixties and seventies, the image of Orchard Beach and Pelham Bay Park has changed again, while Latinos remain the dominant visitor group to the park. Orchard Beach has come to be known as a beach with an animated and sociable atmosphere, a place where families can spread out, hook up with others, enjoy energetic music and dancing, while safely recreating in the park for long hours.

Over the past 20 years, the park has continued to be popular, although the visitor population has fluctuated in number. This may be the result of changes in the overall patterns of recreation as the number of shopping malls and other air-conditioned spaces expanded. This fluctuation may also relate to challenges of managing parks on limited budgets. Maintaining a balance between the visitors' needs and the park's ability, on shoestring budgets and insufficient staffing, to meet those needs has always been a tricky business, as we have seen from the history of Pelham Bay Park. Today's park management also faces the challenge of accommodating new visitors from Central and South America, the Caribbean, eastern Europe, and East Asia and of managing new forms of leisure practices including Jet Skiing, in-line skating, and the often invasive use of boom boxes, powerful car stereos, and camcorder equipment operated by visitors. These challenges must be met while maintenance of the aging park is ongoing. The following discussion illustrates a successful equation of visitor initiatives and management capabilities in the context of diverse users and practices, limited budgets, and the distinct social history of Orchard Beach.

Latino Visitors

As previously mentioned, Orchard Beach is an important site of recreation for an active Latino community. In the summertime the park holds weekend concerts that draw some of the largest crowds that the park sees all year, a significant portion of which are of Spanish-speaking cultural backgrounds. Informal activities, too, draw Latino visitors on a regular basis. On a typical summer day one finds numerous Latino families celebrating birthdays and family reunions on the beach under tents and under the shade in the picnic areas. Many of these gatherings are annual events that intend to bring members together from all corners of the city. Based on face-to-face interviews, Latino visitors come to the park from diverse neighborhoods, including parts of the South Bronx, Washington Heights, East Harlem, and Astoria. Members of these parties bring hammocks for siestas, card tables for dominos, and cookout grills for cooking chicken and chorizo.

At the picnic site families display national flags and decorated banners with family names and sayings in Spanish and English. Couples dance in pairs on the sand or on patches of grass to the sounds of merengue, salsa, or cumbia music emanating from small portable boom boxes. Within these large groups typically two or three generations of family members are present and at one point or another, everyone takes a turn playing with and supervising the children. There is also a good deal of movement between groups as parties borrow from one another minor supplies that they might have forgotten in the rush to get to the park. The exchange of goods and kids sent to ask for them allows for some of the parties eventually to merge together.

In addition to the colorful picnic scene and the organized music program, Latino visitors organize their own informal dance parties on the weekends. During the summer our study took place, a particular visitor named "Frankie"[4] voluntarily set up his own sound system in a section of the activity area.[5] On a weekly basis he became the master of ceremonies to a lively salsa party. Frankie's program enlivened the park with peppy music. He motivated visitors of all ages to pair up and to swing around the dance floor.

On occasion, a parade of Swing Bikers appeared in the park to add to the festivities. The Swing Bikers is a club predominantly made up of individuals who collect sharp-looking antique bicycles dating from the 1950s. Members tie mini

Figure 6.4. Picnicking at Orchard Beach

Puerto Rican flags to the handlebars and slowly spin around the park proudly showing off their nostalgic bikes. On the weekends, between Frankie's show and the parade of Swing Bikers, the park was an animated and picturesque scene, alive with dancing, socializing, music and sights to see. Large crowds gathered around the ball courts to take it all in.

It is uncertain whether Latinos' strong sense of place attachment developed from visitor-initiated events or park-organized programs. What is certain, though, is that the formal and informal events contribute to maintaining as well as developing community identity. The fact that the park management is not in a combative relationship with its users contributes to this development. Park management accommodates the Latino community through a healthy mix of organized events and by honoring visitor-initiated activities with spaces of their own, however temporary and provisional these spaces may be. In these ways, the park managers communicate their knowledge of and respect for visitor groups and their cultural expressions.

The image of the park as a piece of the broader Latino community in New York City has different meanings for different groups of Latino visitors. Some visitors come to the park because of the animated atmosphere and the presence of other Latinos. This view led one "Nuyorican" to comment that "90 percent of the people here (at the park) are Latino . . . that's why we're here." Other Latinos feel that too much Latino music and emphasis on the Latino community generalizes one's cultural identity and overshadows diversity within the community. One visitor complains "about the musical performances . . . before they used to give you options like CD101[6] . . . they'd play different music like jazz. . . . I'd like to have this music again." Visitors identify positively but ambivalently with the strong presence of a Latino community at Orchard Beach. Regardless of this range of reaction, both kinds of commentary confirm identification with a Latino community that is present in the park. Such a sentiment echoes the former Tent Colony, whose members, although not always in agreement with one another, shared the customs, rights, and responsibilities of a common community identity.

Some members of today's Latino community see the symbolic cultural expressions of their heritage as a draw for others who are not necessarily defined as part of the community. One Latina comments, "When there is salsa music, it's beautiful . . . it is from our culture . . . it is beautiful to see diverse nationalities enjoying it." Such a view of a distinct cultural group stands in contrast to past patterns among visitor groups at Orchard Beach. In the past, white ethnic groups operated as cultural "clubs" within the park that excluded nonmembers. One must be born into such clubs, or must "pass" for a member as the light-skinned Italian American mentioned earlier once did in order to socialize with German Americans at Orchard Beach. The open posture of Latino visitors toward the presence of Latino culture in the park is inclusive and encourages diversity.

Local Seniors

Although the park is dominated by Latinos, other cultural groups enjoy it too. Local seniors comprise a distinct and visible group at Orchard Beach during the off seasons and times. On a weekday in the fall one may expect to find an empty park. However, Orchard Beach is typically visited on a regular basis by a noticeable group of approximately 20–25 seniors. Many of the seniors live in local neighborhoods just a short drive away from Orchard Beach, such as Pelham Parkway, Morris Park, Coop City, City Island, and parts of Westchester County.

Over the past 30 years, the Bronx neighborhoods in which local seniors reside have changed demographically. Once inhabited predominantly by Italian American, German American, and Jewish families, these areas have growing proportions of Latino, African American, and Asian American residents. According to census data reported in the *New York Times* (Pierre-Pierre 1993, 31),

Figure 6.5. Seniors at Orchard Beach

in 1980 whites made up 77 percent of the residents of Pelham Parkway. That number fell to 59 percent in 1990, while the proportion of Latino residents rose from 13 percent in 1980 to 24 in 1990. The percentage of African Americans rose from 7 to 12, and the Asian American population from 1 to 3 percent between those same years (Pierre-Pierre 1993, 31). These neighborhoods also are associated with the oldest populations in New York City. In 1990 just over 49 percent of the residents of Pelham Parkway were over 65 (ibid.).

Generally, the local seniors are found in the park during the hours and seasons when there is low visitation, including early mornings in the spring and summer and throughout the day during the fall and winter. On a typical early morning in the park, one may see seniors of a variety of ages engaged in diverse activities. The "in-betweens"—individuals of retirement age "still trying to be young" (as some seniors refer to their peers)—are found in groups on the handball courts. Others stroll along the promenade or on the nature trails in pairs or threesomes. A group of "regulars" sit in the sun in foldout chairs at the Ecology Center. This small building was once a concession stand and now is used to store equipment. On occasion it is used to meet with educational groups touring the park. The Ecology Center is an underutilized resource in the park that the seniors have temporarily claimed as their territory.

The following excerpt from field notes highlights the way seniors have rooted themselves at the Ecology Center and use it as a site within the park to express their cultural identity:

November 7, 1996

As I was walking down the promenade in the direction of section 1, I noticed an American flag flying on the corner of the railing that led up to the Ecology Center. In the summertime I met a constituent who mentioned users who flew an American flag. At the time, I didn't know if he was exaggerating or not. Now I know.

Behind the flag I saw that there was a small group of adults sitting in chairs behind the picnic tables. There were three men and one woman, all white, all appeared to be of the age of retirement. The woman, actually, was lying down on a foldout body-length lawn chair. Together the group was chatting and listening to big band music on a small radio that was balanced on the edge of a windowsill at the Ecology Center.

One interesting aspect of the group is that they see themselves as an exclusive club. One man invited the interviewer to have a look inside the "club house," the term then used for the Ecology Center. Inside a storage closet with outdoor access off of the Ecology Center the seniors have a "kitchen." Here were hanging at least 20 mugs on pegs on a wall and a plug-in coffee pot sat on a countertop. In the kitchen there is a small sink and stocks of coffee and tea supplies. There are numerous photographs pinned to the door, covering practically every inch of it. The members explained that the photographs are of friends with whom they used to socialize at Orchard Beach. Many of the photos appear to have been taken at Pelham, yet others were taken inside private homes. As seniors talked about their friends one by one, it was revealed that many people in the photographs are deceased. Nonetheless, the photographs are talked about with excitement and fond memories.

The association that these local seniors at the Ecology Center have formed extends to other areas of the park as well, and to areas beyond the boundaries of the park itself. Coming to the park on off hours is a strategy local seniors use to achieve a sense of membership within their club. Off hours allow them to have temporary ownership of the park and thus to avoid conflict with other visitors. In part, this strategy may relate to the sense of vulnerability that comes with the aging process. It may also reflect this particular group's vulnerability as a faction of people who have broken off from their larger cultural group. The local seniors comprise a community of people who did not migrate out of the Bronx and now have become a minority cultural group. The Bronx neighborhoods have changed, but the local seniors still consider the borough and its public resources "their" territory to defend and to maintain authority over. One elderly person expressed this sentiment when he said, "This is our home ... there is nothing to be afraid of."

Race and ethnic identities play a part in local seniors' sense of community. Frequently and without solicitation, participants introduced into the conversation disguised commentary on the racial differences of park visitors. One senior characterizes his use of the park during off hours in the following manner: "You are more or less isolated here.... The ethnic groups stay in one place.... I don't bother them ... nobody is bothering me ... why look for trouble?"

When the topic arose, interviewers asked participants to discuss their views on the racial composition of the park. In these instances, seniors displayed discomfort. They used "us" and "them" terminology to skirt around labeling cultural groups even when empathizing with minority populations who are treated unfairly as a result of the power structure in the United States. One senior commented, "I think for *these* people, it should be free to park here ... these are not people from Larchmont[7] ... they just want to get off their fire escapes and beat the heat." All participants in the study were asked how one identifies him or herself racially, ethnically, culturally, or otherwise. Seniors typically avoided this question by supplying wry answers such as, "That doesn't enter into it" and "What is the next question?" In short, what they said and did not say conveyed an anxiety about protecting oneself and territory from others whose racial backgrounds differ from their own.

In some ways local seniors at Orchard Beach are similar to Latino visitors and echo, to a certain degree, the traditions of the Tent Colony. Like these groups, local seniors take initiative in creating their own spaces. An American flag, music, coffee cups, and images of their peers symbolically express their strong attachment to the park, their claims to a space, and their identity as a small though perhaps exclusive group. Local seniors call to mind the Tent Colony more so in that they heavily rely on their relationship with the park management for protection and for validation as a group. The informal and temporary agreement between park managers and local seniors adds to the group's sense of belonging to the park. One informant commented, "We love everything. This is home all year round. They know us, the parks people. They know who we are. We have the flag. We appreciate the flag of the United States and we value the flag and the park." Knowing that the park management "approves" of their use of the park and believing that they are given "special" treatment strengthen the relationships between the management and its constituents and between park users and park space.

The Naturalists

Typically cultural groups are imagined as communities of individuals sharing similar experiences and perspectives associated with a particular heritage, religion, class, gender, sexual orientation, or age. Cultural groups, however, may also emerge among people who share similar experiences through their engagement in particular recreational activities. Green parks offer leisure opportunities that distinctly contrast with those offered in shopping malls, movie theaters, museums, and other indoor cultural institutions. As a result, green spaces give rise to particular cultural groups such as "naturalists"—park visitors whose primary activities include walking, exploring the natural offerings (e.g., trails, the beach, wooded areas), dog walking, fishing, and camping. In the context of the broader city, with few, mostly underfunded green spaces, naturalists are a marginalized cultural group.

The naturalists at Orchard Beach are people from diverse cultural backgrounds. They represent various ages, socioeconomic levels (education and income), and places of birth, although fishermen are an exception. At Orchard Beach, fishermen are predominantly Latino and African American men. In some cases women accompany fishermen, but they rarely engage in fishing.

We found that the naturalists tend to come to the park on a regular basis for routine morning walks and scheduled time to be alone. Fishermen reported coming to the park in time to meet fish migrations, although many visit the park without their equipment to engage in other forms of outdoor recreation.

In general, naturalists are visitors who are in tune with changes occurring in the natural areas. They notice the details of park maintenance and are quick to evaluate the park management's response. One woman commented, "If a tree falls across a path, the parks people are fairly prompt in getting it out of the way." Other naturalists do not have very high expectations for park maintenance. Their point of view echoes the views associated with the original preservation concept of the park. One woman hiker commented, "One of the nice things about the park is that you don't have to do too much. There is solitude. It is easy to walk here. There is access, and there's a healthy variety of walks that you can take." In the view of this visitor, the park is vast and there are many options for walking. Occasional fallen trees and debris blocking the trails do not prevent her from enjoying the park, for she claims she always finds an alternative route. Other naturalists see themselves as responsible for the maintenance of the park. They take initiative in clearing paths, and they rake the leaves out of their favorite areas. Some bring extra trash bags to distribute to other visitors in an effort to keep the park clean. In many cases these visitors engage in these activities to make overt, political statements about protecting the natural environment. Others consider these activities an ordinary aspect of their visitation.

The symbolic expressions of naturalists are slightly different from the expressions of Latinos and local seniors. Naturalists are not generally interested in marking their presence in the landscape by posting cultural flags, playing music, or altering natural areas as a means of personalizing the park. In fact, naturalists commonly share the belief that one's respect for nature is best "marked" by the invisibility of human presence or impact. Much in the way that anthropologists see browsing and reflecting upon objects while shopping as a process whereby consumers transform anonymous commodities into personal possessions (Carrier 1993), ritualistic walks along the beach, forays into the forest, and "hunting and gathering" activities are the means by which naturalists appropriate the park. As naturalists take their routine walks through the landscape, they trace and retrace the landscape within their imaginations. The process of repetitively viewing the landscape becomes a process of identifying with its elements and "reading" oneself into the scenery. Through ritualistic behavior naturalists come to possess the park and to develop deep attachments that are seemingly intangible. The symbolic manifestations of naturalists' attachments, therefore, are most visible in their routines and verbal expressions about the park.

Naturalists from diverse backgrounds discussed their identification with the park's landscape in similar ways. The theme of nostalgia for another place and time is present in many visitors' expressions. The close proximity of the trees to the water, for example, prompts many Latino visitors to reminisce about the shorefronts in the Caribbean and other parts of the Americas. These comments echo those of the local seniors who often suggest that the landscape conjures memories of a time when the city, the park, and their power as an urban cultural group was different.

Some naturalists see the landscape in terms of the experiences of early European explorers viewing the New World for the first time. Many commented to the effect that the park is "wild." It is an untamed place to uncover, conquer, and settle. One visitor described himself as an explorer searching for treasures. In fact, he combs the beach with a Geiger counter, sifting through the sands for rings and other small objects of value that visitors may have left behind. As this visitor explained to a researcher his systematic technique for covering the beach, and his fine-grained observations of the changes in the landscape, it became clear that this visitor is a modern explorer with a scientific approach to discovery.

Another visitor, the daughter of a former colonial administrator in Africa, reports her use of the park with a similar Westerner's spirit for the conquest of nature. She commented:

> I walk three miles on the weekend to picnic. I meet up in the park with a friend at 7:00 a.m. and we bring bagels and muffins and have breakfast in the woods. We go way into the back of Hunter Island. We do this in the middle of winter with Schnapps and hot coffee. The dog has a back pack to carry the things for us. The park is especially beautiful in the winter.

This description was delivered with much excitement, suggesting the park visitor is titillated by the possibility that her use of the park is "out of bounds." Her visitation to "remote" places during off hours and seasons suggests that for her visiting the park is an act of crossing an imaginary boundary between civilized and uncivilized worlds. On the one hand, this visitor's views, especially her deep appreciation for the "pure" beauty of the park, reflect an Emersonian worship of nature. She recounts being accompanied by a domesticated animal on her way to perform a breakfast ritual in the "deep" of the forest.

The illustrative expressions of naturalists and the ways that they imagine their place within the landscape point to an important achievement of the park management. Although the original concept of Orchard Beach once emphasized the value of preserving nature, and over time the emphasis has shifted to organized recreation, urban visitors are still drawn to the powerful landscapes that continue to provide cultural symbolism, inspire creativity, and fuel the imagination. In this light the park stands as an example of a public space that has captivated the imaginations of visitors over a long span of time. It also is representative of public spaces that span the dichotomous definition of a park as either a preservation or a recreation area. In intentional and unintentional ways, the management of Orchard Beach provides visitors with a bit of both.

Conclusions

Taking note of symbolic cultural expressions can be useful to park managers for a number of reasons. First, parks are not neutral spaces. They are socially constructed and they have complex histories. An analysis of the symbolic expressions of diverse cultural groups reflects a park's contemporary history and social climate. Descriptions of cultural groups and their symbolic expressions add depth to an understanding of the social life of a park and how it has changed over time.

Second, varied expressions of cultural pride in the form of flags, family banners, music, and ways of claiming spaces within the park indicate important connections that users are in the process of developing and protecting. In theory, public parks are resources intended for use by all members of society. However, urban public parks do not always reflect the social diversity of a city. Often working-class, poor, elderly, and female citizens and minority cultural groups feel unsafe and out of place in public space. In the case of Orchard Beach, however, Latinos, local seniors, and naturalists—three marginalized cultural groups within the city—have created their niches within the park. Park managers can "read" diverse cultural expressions as indicators of groups' sense of belonging. Knowledge of cultural expressions can inform managers of groups that are becoming detached from the park, or coming into conflict with each other. This knowledge can be used to develop diversity and equality while preventing cultural conflict. It can help managers gauge whether laissez-faire or hands-on management is the more appropriate approach.

Third, an analysis of symbolic expressions can clearly indicate to managers the particular worldviews and practices of visitor groups. Understanding a park's constituents becomes helpful in prioritizing the relevance of resources to the park. Certain cultural manifestations may seem impractical or disrespectful of park rules in park managers' eyes. For example, roping off large picnicking spaces before many other visitors come to the park, drinking alcoholic beverages in the woods, and dancing and playing loud music amidst other visitors, may be troublesome behaviors to managers. Yet, other cultural behaviors may appear silent, invisible, or so ordinary that they are overlooked as insignificant—for example, the rituals and perceptions of naturalists. When examined closely, invisible or ordinary cultural practices are revealed as complex and significant, while attention-calling or "problematic" behaviors are revealed to be based in rational cultural logics. Ultimately, understanding the cultural practices and learning to read constituents' symbolic expressions can enhance managers' awareness of visitor groups and facilitate the prioritization of resources for development.

Finally, cultural knowledge of visitor groups brings to the attention of managers as well as visitors the ways that parks help sustain the cultural dynamics of urban societies. Parks provide numerous resources that cultural groups use for the continuity of their communities. For example, Orchard Beach plays a major role in providing a "home" for Latinos and local seniors who for many reasons feel marginalized and lack spaces in the city that they can call their own. Parks are significant to carrying out and sustaining the processes of cultural reproduction. For this reason, on-the-ground knowledge of cultural groups and their relations with one another illuminates the cultural diversity of cities and, in doing so, underscores the significance of public parks within the urban environment.

Notes

1. Jacob Riis Park has a lively core group of visitors, but in terms of numbers, it is an underutilized park most of the year.

2. The four points for counting include Bruckner Boulevard opposite the Pelham Bay terminal of the No. 6 subway line, Bruckner Boulevard and Wilkinson Avenue, Middletown Road and Rice parking lot, and Orchard Beach at the parking lot.

3. As documented by Catherine Scott (1993), Pelham Bay Park has a long history of accommodating government agencies within the park. Not only did the U.S. Navy occupy Rodman's Neck and Orchard Beach for a time during World War I, but the New York Police Department used Rodman's Neck between 1930 and 1936 as a training ground. The U.S. Army used Rodman's Neck as an antiaircraft gun site in the postwar years until 1956, and from the 1950s to the present the NYPD has used Rodman's Neck for its Firearm and Tactics Unit training.

4. Pseudonyms are used in this chapter to protect the identities of visitors.

5. This party has been taking place at the park for many years. Prior to the year of our study, the party was typically held in a spacious area near the bathhouse. During the time of our study, however, the bathhouse was under renovation. The park management shifted Frankie and his tall speakers to a run-down ball court that did not receive much use. The ball courts were next in line for repairs.

6. CD101 is a popular jazz radio station in New York City.

7. Larchmont is a middle-class to upper-middle-class community in Westchester County, about five miles north of Pelham Bay Park.

Chapter 7
Independence National Historical Park
Recapturing Erased Histories

Introduction
Erased Histories

As I (Setha Low) drive Route 10 from Palm Springs to West Los Angeles, my personal history passes by inscribed in the landscape through places, institutions, and cultural markers. I am reminded of where I went to college, where I spent my summers as a child, and where I got my first job as I travel this Southern California highway. Physical reminders like these provide a sense of place attachment, continuity, and connectedness that we are rarely aware of, but that plays a significant role in our psychological development as individuals and in our "place-identity" or "cultural-identity" as families or ethnic and cultural groups (Low and Altman 1992).

But what happens when your places are not marked, or even more to the point, when your personal or cultural history is erased—removed from the landscape by physical destruction? The redevelopment of Paris by Baron Haussmann and removal of buildings around Notre Dame in the nineteenth century is a classic example of the erasure of a working-class and poor people's history in an urban landscape (Holleran 1998). In the United States we have been more subtle; for instance, the contextually complex, residential streets of Bunker Hill were "lost" in the modernist redevelopment of downtown Los Angeles (Loukaitou-Sideris 1995; Loukaitou-Sideris and Dansbury 1995–96). At the historical site of Independence National Historical Park in Philadelphia, there is no record of the people who built the buildings (African Americans),[1] or who financed the Revolution (Jewish Americans), or who fed the soldiers (women—mothers, wives, and others). The processes of historic preservation, planning and development, and park interpretation recreated the colonial period as a white, male space. Further, documentation of lost buildings and physical context is missing when one searches for information concerning the histories of minority peoples during colonial times.

African Americans in Philadelphia, however, have been fighting to reclaim their history by supporting research, setting up archives, and working to ensure that their history and culturally significant sites are marked throughout the city. The African American community in New York has been successful in contesting the federal government's claims to the African Burial Ground, demanding its commemoration and preservation, but has been less successful when it comes to preserving their political heritage, a case in point being the demolition of the interior of the Audubon Ballroom, where Malcolm X was shot. Thus, even as histories are erased, they are re-searched and rediscovered so that they can be commemorated in the contemporary fabric of the city.

This chapter presents the results of a collaborative research project that uncovered the erased histories of various ethnic groups living in Philadelphia and recaptured these histories through a rapid ethnographic assessment procedures (REAP) study of Independence National Historical Park. It tells the story of how the planning and design of the park over time unintentionally disrupted the cultural attachments of neighboring communities and excluded new immigrant groups. An understanding of the underlying processes that created this disruption and exclusion is illuminated by the application of the REAP.

In the following sections, we review the methodology and discuss findings from the various qualitative methods. In the conclusion of this chapter, we address the importance of this type of research for understanding the impact of design and planning on the place attachment and cultural identity of local minority populations. We offer this case study as representative of the kind of research that can be undertaken to recover histories that have been scrubbed away during previous historical preservation or urban renewal projects. Further, we argue that understanding the intimate relationship between histories, cultural representation, and park use is critical to successful design and planning in any culturally diverse context.

One note about our use of terms. Anthropologists argue continuously about the categories of ethnicity and culture. *Ethnicity* is a slippery term that evokes different meanings when used by the informant as an identity marker ("I am ethnically Jewish," or "I am Italian American") and when used as an analytic category ("The informant appears to be Asian American or African American"). Culture is equally difficult to use. The term *culture* refers to local traditions or practices that define an ethnic group, while *Culture* refers to an analytic category with overarching, anthropological meaning. Further, ethnicity and culture covary with nationality and other political forms of identification. In this chapter we use ethnic group and cultural group interchangeably, without untangling their multiple intellectual histories. We prefer the term *cultural group* as it most closely reflects the traditions and histories discussed. On the other hand, Italian Americans and Puerto Ricans are considered ethnic groups in the sense that ethnicity is colloquially understood as an inheritance and/or determined by immigrant-group status within the United States.

Methodology

In 1994 Independence National Historical Park began developing a general management plan that would set forth basic management philosophy and provide strategies for addressing issues and objectives over the next 10- to 15-year period. The planning process consisted of extensive public participation, including a series of public meetings, televised town meetings, community tours, and planning workshops. As part of this community outreach effort, the park wanted to work cooperatively with local communities to find ways to interpret their diverse cultural heritages within the park's portrayal of the American experience. We designed this ethnographic study to provide a general overview of park-associated cultural groups, including an analysis of their concerns and the identification of cultural and natural resources used by and/or culturally meaningful to the various groups.

Cultural Groups and Neighborhoods

Park-associated groups including African Americans, Jewish Americans, and Italian Americans, whose recent ancestors previously lived in the general area, were identified as the initial groups for contact. These groups were selected because the area has had special importance for them. Other cultural groups such as Asian Americans and Hispanic Americans were included because they were identified as rapidly growing communities who use the park grounds for ceremonial and recreational purposes, and thus would be affected by the proposed changes to Independence National Historical Park.

Four local "ethnic" neighborhoods were selected for study—Southwark for African Americans, Little Saigon for Asian Americans, the Italian Market area for Italian Americans, and Norris Square for Hispanic Americans—based on the following criteria: 1) they were within walking distance from the park (excluding Norris Square); 2) they had visible spatial and social integrity; and 3) there were culturally targeted stores, restaurants, religious organizations, and social services available to residents, reinforcing their cultural identity. We selected the Vietnamese American community to represent the Asian American cultural group because of its proximity to the park and its recent population growth.

We attempted to interview across class in these neighborhoods. Where this was not possible, such as among African Americans in Southwark, we turned to local churches—Mother Bethel in the Society Hill neighborhood next to the park and Nazareth Baptist Church in Southwark—to gain class diversity not present in the Southwark projects. The Jewish community could not be identified with a spatial community in the downtown area; therefore, we decided to interview members of both Conservative and Orthodox synagogues in the Society Hill area as a "community of interest" rather than a physically integrated area.

Rapid Ethnographic Assessment Procedure Methods

A number of methods were selected from the REAP methodology to generate data from diverse sources that could then be integrated to provide a comprehensive analysis of the site. Behavioral maps recorded people and their activities located in the park throughout the day and early evenings on weekdays and weekends. Transect walks recorded what identified community members described and commented upon during guided walks across the site. Individual interviews based on the study questions were completed in Spanish, English, or Vietnamese. A total of 19 interviews for the African American community were collected in the Southwark area and from members of the congregations at Mother Bethel and Nazareth Baptist churches. Seventeen interviews were collected from Hispanic Americans (mostly Puerto Ricans) at the Puerto Rican Day Parade in Independence National Historical Park. Seven individual interviews were collected from Jewish Americans who attended the Society Hill Synagogue. Nine interviews were collected from Vietnamese Americans living in Little Saigon and from members of a Vietnamese American Catholic church in South Philadelphia. The majority of the 19 interviews with Italian Americans were collected outside of Catholic churches in the Italian Market neighborhood.

Expert interviews were collected from individuals such as religious leaders, local historians, historic preservation specialists, and tour guides identified as having special expertise to comment on the cultural significance of Independence National Historical Park. We collected nine expert interviews from the local communities: three from the African American community, one from the Asian American community, one from the Hispanic American community, two from the Italian American community, and one from the Jewish community. A folklorist who specializes in Philadelphia neighborhoods was also interviewed, and a number of local experts were consulted informally.

Focus groups composed of 5 to 13 individuals each were set up with major religious institutions in the neighborhoods—churches and synagogues—as well as with active community organizations such as Asian Americans United and *Motivos*. Each researcher kept a field journal that recorded observations and impressions of everyday life in the park. Historical and archival work accompanied all phases of the study, and newspaper clippings, articles in local magazines, and other media-generated materials were collected.

The data were organized by coding all responses from the interviews, and then analyzed for content by cultural/ethnic group and study question. The transect walks, tours, and interviews were used to produce cultural resource maps for each group. A base map was used to record the existing conditions on the site. Focus groups enabled us to determine the extent of cultural knowledge in the community and to identify the areas of conflict and disagreement within the community. The combination of mapping, transect walks, individual and expert interviews, and focus groups gave us independent bodies of data that could be compared and contrasted. As in all ethnographic research, interviews, observations, and field notes, as well as knowledge of the cultural group patterns and local politics, were used to interpret the data collected. Table 7.1 summarizes the methods, data collected, time frame for data collection, and what was learned from each of the methods.

The idea of a national historical park in Philadelphia originated with the Federal Historic Sites Act of 1935, which authorized the National Park Service to engage in research and educational and service programs and to protect, preserve, and maintain historic buildings and sites for public use. In the late 1940s, the Philadelphia National Shrines Park Commission was created by Congress to plan for Independence National Historical Park, and Judge Edwin O. Lewis of the Philadelphia County Court of Common Pleas was chosen to head it. Planning and site acquisition began in the late 1940s; demolition, site preparation, and construction took place throughout the 1950s.

Custody of Independence Hall itself was transferred from the city of Philadelphia to the National Park Service on January 1, 1951. The Independence Mall area was a project of the Commonwealth of Pennsylvania, while work on the blocks between Independence Square and Second Street was carried out from the beginning by the National Park Service. Planning continued through the 1950s, carried out by an NPS-designated task force that met regularly with community and political leaders. From the beginning, the task force faced several divisive planning issues centering around differing approaches to historic preservation.

One important disagreement was between those who envisioned a formal layout, in which the historic structures were connected to a central axis, and an informal layout that replicated the historic patterns of internal alleys and lesser streets that had evolved over time within Philadelphia's large, square blocks. Judge Lewis, an advocate of the formal, axial concept, had his way with the Mall north of Independence Hall, but lost out in the eastern blocks to the National

Table 7.1. Independence National Historical Park: Methods, Data, Duration, Products, and What Can Be Learned

Method	Data	Duration	Product	What Can Be Learned
Behavioral Mapping	Time/space maps of sites, field notes	2 days	Description of daily activities on-site	Identifies cultural activities on-site
Transect Walks	Transcribed interviews and consultant's map of site, field notes	6 days	Description of site from community member's point of view	Community-centered understanding of the site; local meaning; identification of sacred places
Individual Interviews	Interview sheets, field notes	12 days	Description of responses of the cultural groups	Community responses and interest in the park
Expert Interviews	In-depth interview transcriptions	10 days	Description of responses of local institutions and community leaders	Community leaders' interest in park planning process
Formal/Informal Discussions; Participant Observation	Field notes, interview sheets	20 days	Description of the context and history of the project; description of park needs	Provides context for study and identifies NPS and community concerns
Historical Documents	Newspaper clippings, field notes, collection of books and articles, reading notes	7 days	History of the park's relationship to the surrounding communities	Provides historical context for current study and planning process
Focus Groups	Field notes, tape recordings	6 days	Description of issues that emerge in small group discussion	Enables understanding of conflicts and disagreements within the cultural group

Park Service professionals, led by Charles Peterson, who preferred a spatial organization springing from the historic public ways of passage.

A second major controversy centered around how much of the existing city fabric was to be removed in this effort to create a setting for the eighteenth-century structures associated with American independence. The blocks both east and north of Independence Square were densely built up with commercial structures of granite, marble, and brownstone, built at various times, of various heights and architectural finishes, according to the typical lot subdivision practices familiar in American cities. By the 1950s, most of these buildings were anywhere from 40 to 100 years old, the bulk having been built between 1860 and 1890. In the 1950s climate of slum clearance and urban renewal, such blocks of buildings were regarded by persons such as Judge Lewis, who represented political, civic, and business interests, as symbols of decline and even threats to the continued prosperity of Center City.

On the other side, Peterson and other architects and historians were advocating an architecturally sensitive approach to renewal, one that, while still clearing away most of the nineteenth-century buildings, would leave intact some of the more distinguished structures—distinguished both in their architecture and in their being linked to the area's history as the banking and financial center of Philadelphia in the mid- to late nineteenth century. The Guaranty Building, for example, which stood on Chestnut Street near Carpenters' Hall, was an important late-nineteenth-century bank, designed by the gifted architect Frank Furness. However, Peterson was unable to persuade the decision makers to save any of these nineteenth-century buildings within park territory. Ultimately, the park planners held to the historic preservation philosophy exemplified by Colonial Williamsburg, which creates a seamless, single moment in history—or, in Ada Louise Huxtable's (1997) term, an "unreal" depiction of the site.

The historical park project became part of a larger effort to renew Center City Philadelphia, an agenda conceived by city planner Edmund Bacon and advocated by the Dilworth and Clark mayoral administrations. One focus of this effort was Society Hill, an area east of Eighth Street and south of Walnut, and adjacent to the new national park. The name comes from the eighteenth-century Free Society of Traders, which purchased the land around Dock Creek from William Penn, and the area began as a wealthy residential neighborhood (Warner 1968). By the early twentieth century, however, the wealthy were long gone and the name "Society Hill" had been forgotten. The neighborhood, close to the Philadelphia immigration center on the Delaware at Washington Avenue, had become home to a multiethnic population that included African Americans (this area, although already multiethnic, was part of the area of African American settlement studied by W. E. B. Du Bois in the 1890s), eastern European Jews, Italians, Polish, Irish, and Ukrainians. By the 1940s, the neighborhood had become increasingly poor, as well as increasingly African American, although pockets of the other ethnic populations remained.

Because of the proximity of this neighborhood to the projected park area as well as to downtown, and the high quality of much of its building stock, the city saw the national park planning process as an opportunity to restore the neighborhood to its colonial-period status as an upscale residential area. Edmund Bacon argued before the City Planning Commission in 1947 for the whole area south of Walnut Street to Lombard Street to be included within the national park (Greiff 1987, 52). It was later decided that the city should pursue the restoration of the area in cooperation with the National Park Service; eventually the development process included, along with the City Planning Commission, the Old Philadelphia Development Corporation (a private, pro-redevelopment group) and the Philadelphia Redevelopment Authority.

Whole sections of the original Society Hill area were designated as redevelopment areas. Homeowners could restore their properties according to strict historic preservation guidelines adopted by the planning authorities or sell them to the redevelopment authority. Since few could afford the costly work of historic restoration, most sold out. The city then offered the properties for sale for a nominal price to buyers who could prove that they had the financial resources to restore them. The banks, the real estate community, and the news media cooperated with the city in creating a favorable image of the redevelopment area, thereby creating a market for affluent, mostly white buyers. Thus, over a period of roughly 15 years, the predominantly poor, heterogeneous community of long standing was dispersed and replaced by a new community of predominantly white professionals. This process of gentrification is well documented by Zukin (1991) and Smith (1996) in their political and economic analyses of uneven development in New York and other East Coast cities. In these discussions, the development of the Society Hill area is presented as an exemplar of these transformative processes producing restored, affluent tourist spaces in the center of the city.

The National Park Service accelerated the gentrification process in Society Hill by simultaneously transforming the blocks from the east side of Independence Square to Second Street into a national historical park. Meanwhile, the state acquired all the property within a three-block area north of Independence Hall for the Mall. The formal, axial design preferred by Judge Lewis left no trace of the preexisting city fabric: all the intrablock alleys were removed and the buildings demolished.

The social and physical upheavals involved in creating Independence National Historical Park did little to foster communication with local communities. The extensive demolition removed many of the settings for life, play, and work that had meaning for members of local communities. The national park became a new, artificially created environment from which most signs of the city's ethnic history—its nineteenth- and early-twentieth-century history—were carefully scrubbed away. In particular, the uprooting of the historic African American community from the area surrounding the park is a legacy that influences any effort to build a strong relationship between the park and that community.

Today the park attracts mostly tourists with minimal use by local Philadelphians. Except for special events such as the Puerto Rican Day Parade, which draws a large Latino crowd, or at lunchtime on a sunny day when office workers from nearby buildings use the Mall to eat, the park is underutilized as a recreational resource. Some Hispanic Americans who were interviewed said that they visit the park to see the flowers, or to take a walk. But on most weekdays, the park is empty until tour buses arrive full of lively seniors, school groups, and foreign visitors.

Findings

African Americans

The community of Southwark, on the southern boundary of the Society Hill neighborhood of Independence National Historical Park, is a racially and culturally mixed, low- to upper-middle-income residential area consisting of brick row houses and public housing high rises. The geographical boundaries of the predominantly African American core extend roughly from Queen Street to Washington Avenue between Third and Fifth streets. The core area is a remnant of the African American community discussed above in the historical

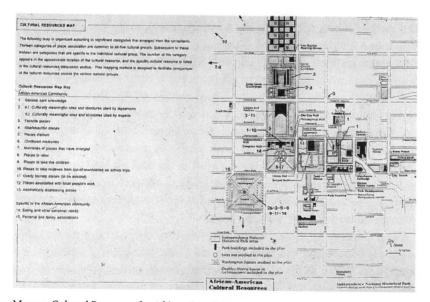

Map 7.1. Cultural Resources for African Americans

section, and one of the few remaining integrated residential areas that has remained in the control of the African American community. Because the majority of Southwark project residents live in low-income households, we selected Mother Bethel Church in Society Hill as a second site for interviewing.

VISITOR USE

Most interviewees acquired knowledge of and association with the park from actual visitation. For example, a couple of the older women commented on having taken their children to the park many years ago, as stated previously. In addition, some interviewees commented on previous visits to the park many years ago, but now they are just "too busy." One man said that he had not visited the park in two or three years, but "the scenery is beautiful. I used to walk and read in Washington Square, and I've been to services at Mother Bethel."

Other interviewees were more negative; one man asserted, "I don't visit [the park] because it has nothing to do for or with me . . . it doesn't show black history or culture. It doesn't represent us and we helped build this country." A middle-aged woman said, "They have to offer us more of everything to get us to go. It's within walking distance but it's not the park I go to. They need to offer us coupons or something to get us there. I go to a park in the neighborhood."

MEANINGS AND SYMBOLS

Of those consultants who felt that the park had meaning, the majority felt that the park was about history and cultural identity. A woman pointed out: "Yes, it's history . . . a part of teaching about history. Some people have lived here all of their lives and haven't visited. There's lots of history here that people don't know about . . . black history." The interview with Charles Blockson, a local historian, reinforced this point. His research reveals that there was a free as well as an enslaved population of African Americans living in Philadelphia during the initial construction of the area that is now Independence Park. He feels that African Americans should have a strong cultural identification with the park because of African American participation in its construction: "African Americans were involved from the inception of the park . . . although we were considered three-fifths of a person, most of them, slave labor, and free African Americans—carpenters, laborers—helped to build, create Independence Hall. We must tell their history" (Blockson 1992).

Other African American consultants felt that the meaning of the park was in the Liberty Bell and the Continental Congress, or in the experience of being there. Pastor Leath of Mother Bethel Church commented that he has taken a carriage ride but has not had a chance to tour the entire park. He said that he had not been to Bishop White's house: "Bishop White was instrumental in the building of Mother Bethel. What is significant about the park is . . . the park is a park. It is blended into the community." Washington Square, which was the original burying ground for African Americans, and later was used as a gathering place, sometimes referred to as "Bongo Park," has meaning for local residents, but it is not commemorated or officially included in the park at this time (see map 7.1).

Many African American consultants, however, said that the park had no particular meaning for them or held negative meanings in that resources would be used for the park, not to help local communities. One elderly woman said: "No special feeling . . . I'm not gonna tell no lies . . . when the kids were small [it was important]." A middle-aged woman commented: "No special feeling . . . the Bell is cracked. . . . What's to see? . . . it's cracked . . . We all know that." A younger woman argued,

> The Liberty Bell, I don't care. We won't benefit at all. . . . It's not beneficial to the people, we don't have any money. When something up there needs fixing up, they come down here to ask about it, what about fixing up something down here? The park ain't done nothing for the people here. This part of town is in isolation. This is not a location on the map. . . . I took my kids to Penn's Landing, and we were looking at the map. We're not on the map. This place is no place.

A man commented that the area had no meaning because "the area is for tourists. It is a white area, the intention is for white people to see the Bell. It is not important for African Americans visiting, it's not for African Americans." He added, "The only thing black at the Park is the ink used." Another consultant said of the park, "It's not important for African Americans, that's why there are no African Americans visiting." A third interviewee stated, "Most people who go there go to look at their own people. It's a showcase for white people."

CULTURAL REPRESENTATION

The overall consensus was that the park does not adequately represent African Americans. Moreover, many responses from both men and women (individuals, focus group members, and expert consultants) were centered around the dearth of diversity in the park, with reference not only to African Americans, but to other cultural groups as well. The majority of the African Americans consulted know that there is a history of African Americans that is associated with the park, they just don't know what it is. One woman in the focus group commented that "you don't see too many African Americans there, just the ones who are working for the park, or outside employed."

This feeling that African Americans are not culturally represented seems to have a long history. One woman in a focus group said that in the past "not everyone was welcome, not everyone was allowed in the park. Now it's different, [but back then] some of our children were not allowed to go into the park. They knew they lived on the wrong side of the neighborhood and that they could not go in . . . they couldn't have picnics there. We didn't go." In the focus group the leader then asked, "Was there something in particular that made you feel you couldn't go there?" The woman responded: "No, not really. You just knew you weren't allowed. If you were there, you weren't there for the right reason." Even today, Pastor Leath of Mother Bethel comments that "there is not a clear message presented in the park. Diversity is not displayed. One could walk through the park and not know that there were African Americans in colonial Philadelphia." Focus group members at Nazareth Baptist Church remarked on the need for different cultures to be able to "grasp" on to an identity: "Different cultures need something to grab, the children need something to grasp at." The speaker felt that there is not a solid identity expressed by the park for people to relate to. A woman added: "An Asian child knows who the Emperor is . . . they know but our kids don't know. . . . A lot has been lost . . . but part of it's our fault too, I guess."

Many of the individuals who participated in the focus groups wished that the park were more relevant in terms of their cultural history. Others in the group felt that cultural history starts "here with the church and the committee . . . the church is the root for us as African Americans." Another man added, "At one time the church was the only platform for people to stand on." But the majority of the people who felt that the park could be relevant agreed with an interviewee who said, "Every child should learn about how we got our freedom," and the mandate to the park is to make colonial history relevant to the African American story.

The majority of African Americans consulted did not make any cultural connections to the Liberty Bell. However, expert educator Charles Blockson (1992) revealed that members of the abolitionist movement previously had made connections. There was a series of books entitled *The Liberty Bell: Friends of Freedom*, written from 1839 to 1859 by abolitionists that refers to the Liberty Bell, and the book series has the Liberty Bell on the cover.

Asian Americans

The data collected from the Asian American community came from Little Saigon and South Philadelphia. Little Saigon is located eight blocks south of Independence National Historical Park along Eighth Street between Christian

INDEPENDENCE NATIONAL HISTORICAL PARK

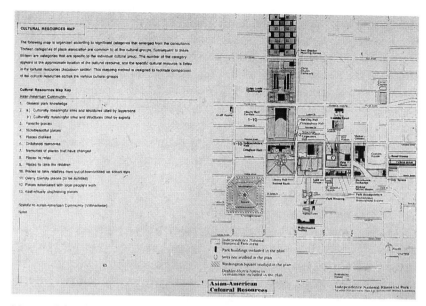

Map 7.2. Cultural Resources for Asian Americans

Street and Washington Avenue. Other small, dispersed pockets of Asian Americans were found between Little Saigon and the Saint Thomas Aquinas Church located at Seventeenth and Morris streets. Finally, an extensive transect walk was taken throughout the Chinatown neighborhood adjacent to the park.

VISITOR USE

The Liberty Bell and Independence Hall were the primary sites of visitation. Interviewees indicated that visits only occur on special occasions such as taking visitors from out of town on a tour, or a rendezvous with a boyfriend or girlfriend. No one mentioned going to the park frequently. Many people said that it was rare to see Asian Americans from the Philadelphia communities in the park.

MEANINGS AND SYMBOLS

Adults, in general, thought of the park as clean, organized, safe, peaceful, and important to the city. Some thought of the park as too far from Little Saigon to visit. Young adults thought the park was "boring" and too "businesslike." Young people also complained that there were no affordable food concessions at the Bourse or in the surrounding neighborhood.

For many, though, the Liberty Bell—particularly its crack—symbolizes the struggle for freedom and rights. The "broken bell" is thought of as attesting to the tensions of those struggles, as well as to the time it takes to find resolution; it is said to represent the "reality" of struggle that should not be hidden. Consultants indicated that these interpretations are rooted in the long, historic struggle of the Vietnamese against the Chinese, the French, and most recently, of South Vietnamese against the North's Communist government. The Bell's crack seems particularly poignant because consultants say that in Vietnam they would never honor an ironwork with such an imperfection. There was some suggestion that the Europeans who made the Bell, though they benefited from Chinese metallurgy techniques, never quite mastered the "recipe." Thus, the Liberty Bell is broken today.

CULTURAL REPRESENTATION

Vietnamese Americans did not necessarily expect to be represented within the park structures, exhibits, or tours. However, consultants did perceive their culture and history to be represented, as previously discussed, in the symbolism of the Bell. Consultants, describing their community members as people who are "eager for education," related Independence Hall and the entire educational mission of the park to the high regard in the Vietnamese American community for education and the pursuit of knowledge, as well as their respect for the role of the teacher.

Several adult refugees suggested that concepts of cultural representation to be conveyed to others might focus on the importance of Vietnamese identity and culture as distinct but not separate from American culture. One community leader stated, "I want to keep the good side of the American culture and the Vietnamese culture.... I want to contribute something here as a refugee. ... I want to be a good person." Younger consultants commented that many immigrants feel conflicted by their transnational ties, responsibilities, and aspirations. They also felt that there are significant generation gaps in the goals and objectives of Vietnamese immigrants. Learning English is seen by both older and younger consultants as central to opportunity in the United States; maintaining the Vietnamese language, though seen by both groups as important in order to retain cultural links, was more important to older interviewees.

In general, interviewees feel that important values and customs are expressed in cultural celebrations. The Vietnamese New Year festivities, in particular, communicate the importance of honoring family, friendship, teachers, and ancestors. This celebration is especially important in the cultural education of children, and in the demonstration of gratitude to the elderly who have a major role in caring for the young. Consultants felt the New Year's celebration is an event the National Park Service might incorporate into their programming as part of an effort to celebrate diverse cultures.

INDEPENDENCE NATIONAL HISTORICAL PARK

In contrast, a Chinatown consultant felt it was important for people to understand the role Asian Americans have played in the general economy and the development of Philadelphia. The mural protest slogan "Our forefathers built the railroad but they never thought it would come this far!" is an ironic allusion to the community's historical participation in national development, as contrasted with the injustices perceived by Chinatown members as a result of rail development in Philadelphia (this slogan refers specifically to the encroachment of the commuter rail tunnel in Chinatown). Members of Chinatown are angry that there is no recognition of their contributions to the city and the important role they currently play in the tourist economy. The consultant argued that the Asian American community of Chinatown forms an important cultural community, which assists the city's promotion of tourism but receives none of the benefits, or any development support from the city. Community members are outraged by the numbers of dislocated low-income families resulting from downtown gentrification projects.

Hispanic Americans

The data collected from the Hispanic American community came from interviews conducted in the park at the Puerto Rican Day Parade, and from a focus group at the Norris Square Neighborhood Project. "Hispanic Americans" is

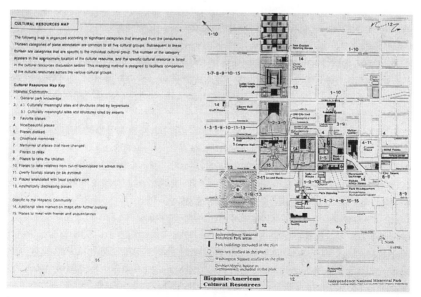

Map 7.3. Cultural Resources for Hispanic Americans

used here as a category inclusive of all Spanish-speaking peoples. In Philadelphia the majority of Spanish-speakers were originally from Puerto Rico.

VISITOR USE

Some interviewees had daily contact with the park as they passed through it on the way to work, ate lunches on the benches in the summertime, and met friends or made business deals in Independence Plaza (see map 7.3). Others have more familiarity with the historical sites through frequent visits with school groups, taking children to play in the afternoon, and bringing visitors from out of town on a tour of the area. One man described his first tour of the park given to him by his cousins as a rite of passage welcoming him to Philadelphia, his new "home." Several interviewees see the park as a place to go with the family, or boyfriend or girlfriend. Many interviewed said they came to the park yearly to attend the Puerto Rican Day Parade while others said that they never go to the park because it is a "museum for (out-of-town) visitors."

MEANINGS AND SYMBOLS

Many interviewees think of the park as a quiet place to relax, with pretty flowers, friendly park rangers, and tight security. The "wooden" benches, brick walkways, and water fountains have been complimented for keeping that "Old Philly" look. It is a place for the family to feel united. One person said that the architectural and design integrity of the park within the city was very important. Many commented that they liked the people dressed in colonial costumes. Another interviewee stated that she gets goose bumps when entering the room where the Declaration of Independence was signed. Others reacted to the park by calling it too "serious." The problem with the park was that: "Quaker history is a dull history." The Liberty Bell was seen as not "kid-friendly." One interviewee said, "The park is a tourist place. It's not an unfriendly park, but it's not inviting either."

A few consultants said they believe that the park is an important place in United States history. To some consultants the park represented the right to freedom and the struggle and sacrifices of the "brave and proud" Founding Fathers. One person commented that the message of freedom was important because "we sometimes take freedom for granted."

"If it weren't for the Declaration of Independence," an elderly interviewee commented, "then the British would still have control of us." Some Hispanic Americans link the symbolism of the Liberty Bell to the colonial struggle for freedom by Latin American countries. Others stated that the freedom message was not a particularly Puerto Rican message. One man added, "A lot of people don't see us as American citizens, as participating in the country. . . . I don't see the connection of our group to the park. I just see the people behind the scenes. ... I think of what goes on that we can't see."

CULTURAL REPRESENTATION

Many interviewees recognized the Puerto Rican Day Parade as a representation of Puerto Rican culture. One person commented, "If you're Puerto Rican and you don't know about the parade, then you're not Puerto Rican. It's in the blood!" However, many criticized the mixed messages this parade sends to Puerto Ricans about their heritage, and to Philadelphians about Puerto Rico. A few consultants were particularly critical of the cultural authenticity presented by the parade. They argued that certain groups who participate didn't play a role in Puerto Rican history. Some saw a need to better utilize the park space for the finale of the parade. They commented that there was too much dispersal. For example, the park might be used to collect people around cultural exhibits after the parade. Other interviewees saw the parade as the only time the white and Hispanic communities came together in the Society Hill area. Still other Norris Square representatives felt that Puerto Rican celebrations shouldn't be focused in neighborhoods where Puerto Ricans don't live.

Some interviewees believe that if freedom and liberty are to be the main messages of the park, park administrators should deal with the issue in terms of all communities that struggle. One man said, "We should all be treated equally; just because we are Puerto Ricans or blacks shouldn't mean anything. There should be no specific place or preference given to other groups." The counter opinion surfaced as well: "We see exhibits about their culture; they should see ours too." Finally, some Puerto Ricans doubted that the park had the administrative and budgetary power to send out any new messages.

Italian Americans

We focused on the historically Italian American neighborhood around the Italian Market, which extends along South Ninth Street between Bainbridge Street on the north and Federal Street on the south. The area on either side of the market, between Sixth and Eleventh streets, continues to have a substantial Italian American population, but with the arrival of Vietnamese immigrants in recent years, it is no longer solely an Italian American stronghold.

VISITOR USE

Many consultants made a point of saying they never go to Independence National Historical Park or that they went when they were kids. One man in his seventies said, "I lived there when I was a kid." Another added, "I went there all

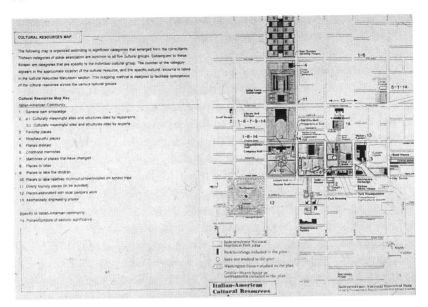

Map 7.4. Cultural Resources for Italian Americans

the time when I was young; I never go there now." Several people mentioned danger—"The town's not so safe anymore"; "I almost got mugged up there"—although some still made the trip sometimes, if not specifically to visit the park. Others said they take visiting relatives there. One retired man had wanted to visit the Liberty Bell when in the neighborhood once, but didn't because of the crowds. A 19-year-old woman explained why she goes to the park: "I like the scenery, and it's a good place to think." Three women had worked in the neighborhood. A 27-year-old woman thought other parts of town more interesting: "Overall, there's nothing there that appeals to me. My sense of Philadelphia history isn't that. . . . I'm more interested in the upper part of Center City, not the colonial."

MEANINGS AND SYMBOLS

The transect walker had a personal relationship with the places in the park. Among the many places with meaning for her were *(a)* Carpenters' Hall, where the members were like a medieval guild who "felt strongly enough about what they wanted [for] their community that they felt compelled to assemble"; *(b)* Old City east of the Mall, with its artists and galleries: "active, thinking people . . . that are vital to the continuation of culture, but also vital to the continuation of the freedom of the country"; *(c)* the open space around Dock Creek, which to her "symbolizes the spaciousness of the Colonial city"; *(d)* the "inviting corridors" created in parts of the park where "direct visual connections enable symbolic associations"; *(e)* Christ Church, where by sitting in Washington's pew she can "realize that these men had lives ... they were people, and they were worried about how they were doing, and they had to pray to God and ask for help too"; and *(f)* the Mall, which she called a "monument to freedom."

Other interviewees cited nonspatial meanings: focus group members cited the Declaration of Independence and the Constitution as creating the setting for the immigration, community making, and prosperity of their ancestors. One man at Pat's Steaks, a renowned fast food hangout in South Philadelphia, and some of the men in the focus groups knew of an Italian American signer of the Declaration, William Paca.

CULTURAL REPRESENTATION

Members of the focus group cited music, the arts, and "the arts of manual labor"—noting the stonemasons and bricklayers, who are "artists in their own right"—as important cultural features. They also felt that the Balch Institute capably portrays the history of immigrant groups, especially the Italians. They suggested collaborative work between the park and the institute, saying, "You don't need to reinvent the wheel." The transect walker felt the park should continue to concentrate on the 1776–1800 period. One woman liked the availability of ethnic community information at the Visitor Center. One woman said, "The park should be for everyone. There are too many nationalities [to emphasize individual ones]." The service station operator agreed: "That causes problems—everything's so mixed. If you have too much of one, it displeases someone else." Four retirees lounging at Pat's Steaks immediately identified with Italian American food: "Pretzels, Italian water ice, hoagies, pizza, apple on the stick—they sell it all around there. Everybody eats Italian food."

None of the consultants except the transect walker felt that the park was relevant to them. One man in the focus group believed that Italian Americans were more clustered in their communities than other cultural groups and were "disconnected" from American governmental and economic institutions. According to them, the story of Independence National Historical Park was a "WASP [white Anglo-Saxon Protestant]" upper-class tale remote from the Italian American Catholic experience.

Jewish Americans

South Street and the streets to either side of it, between Spruce and Catharine streets, approximately, and from Front Street west to about Eighth Street, was originally settled by the eastern European Jews who immigrated between 1880 and 1920. The area became predominantly Jewish by 1910 with a large number

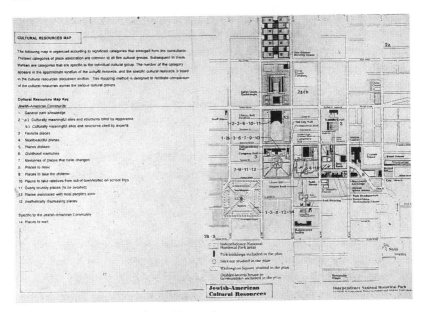

Map 7.5. Cultural Resources for Jewish Americans

of synagogues. What is now the Society Hill Synagogue on Spruce Street was a Baptist church until the Roumanian-American Hebrew Congregation, a major Center City Orthodox congregation, acquired the building in 1910. The Roumanian American synagogue existed until 1966, by which time much of the Jewish population had moved out of Center City. With increasing assimilation and prosperity, Jewish Americans moved out of the ghetto and generally northward, first to North Philadelphia and more recently to the Northeast and to such northern suburbs as Elkins Park. In 1967 the former Roumanian American sanctuary was acquired by the Society Hill Synagogue, a newly formed congregation of relatively affluent Jews, many of whom participated in the renewal of Society Hill then under way.

VISITOR USE

One woman remembered that when the Liberty Bell was in Independence Hall, she could touch it and enjoy feeling its smooth surface. It shone from the touch of so many children's hands. She feels visiting the Bell was better then because it was a hands-on experience. Now, the Bell is in a "cage." Another woman who lives and walks regularly in the area thinks about the difference in use between Washington Square and Independence Square—in the latter, she says, there are "ten times as many people as in Washington Square, and all speaking foreign languages."

We have added this enhanced insert of the cultural resources maps to improve their readability, to facilitate their use as tools for planning and managing the cultural resources in urban parks and heritage sites, and to communicate important cultural information for park users.

CULTURAL RESOURCES MAP

The following map is organized according to significant categories that emerged from the consultants. Thirteen categories of place association are common to all five cultural groups. Subsequent to these thirteen are categories that are specific to the individual cultural group. The number of the category appears in the approximate location of the cultural resource, and the specific cultural resource is listed in the cultural resources discussion section. This mapping method is designed to facilitate comparison of the cultural resources across the various cultural groups.

Cultural Resources Map Key
African American Community

1. General park knowledge
2. a) Culturally meaningful sites and structures cited by laypersons
   b) Culturally meaningful sites and structures cited by experts
3. Favorite places
4. Nice/beautiful places
5. Places disliked
6. Childhood memories
7. Memories of places that have changed
8. Places to relax
9. Places to take the children
10. Places to take relatives from out-of-town/visited on school trips
11. Overly touristy places (to be avoided)
12. Places associated with local people's work
13. Aesthetically displeasing places

Specific to African American Community
14. Eating and other personal needs
15. Personal and family associations

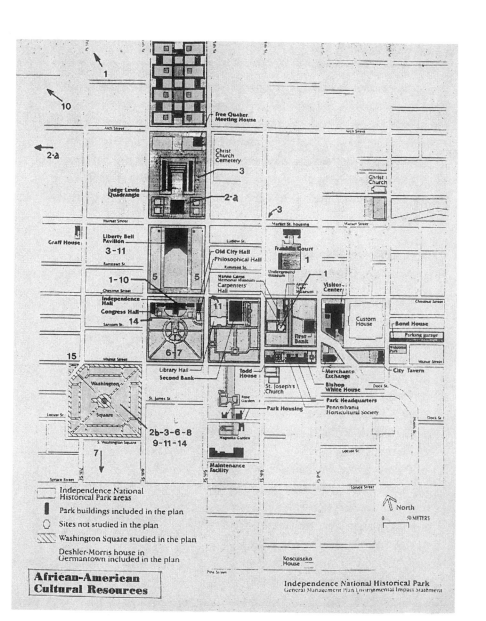

CULTURAL RESOURCES MAP

The following map is organized according to significant categories that emerged from the consultants. Thirteen categories of place association are common to all five cultural groups. Subsequent to these thirteen are categories that are specific to the individual cultural group. The number of the category appears in the approximate location of the cultural resource, and the specific cultural resource is listed in the cultural resources discussion section. This mapping method is designed to facilitate comparison of the cultural resources across the various cultural groups.

Cultural Resources Map Key
Asian American Community

1. General park knowledge
2. a) Culturally meaningful sites and structures cited by laypersons
   b) Culturally meaningful sites and structures cited by experts
3. Favorite places
4. Nice/beautiful places
5. Places disliked
6. Childhood memories
7. Memories of places that have changed
8. Places to relax
9. Places to take the children
10. Places to take relatives from out-of-town/visited on school trips
11. Overly touristy places (to be avoided)
12. Places associated with local people's work
13. Aesthetically displeasing places

Specific to Asian American Community
   None

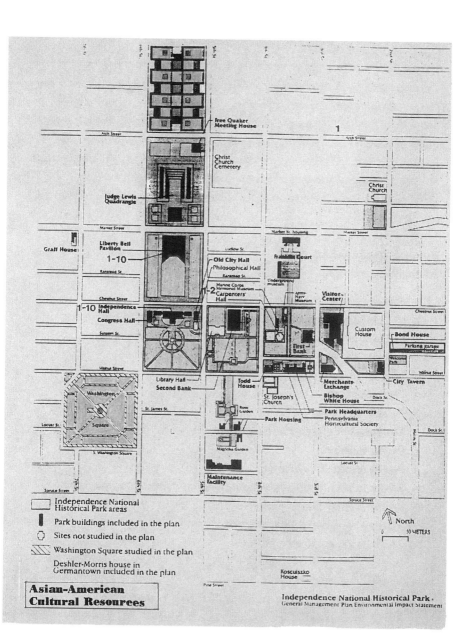

CULTURAL RESOURCES MAP

The following map is organized according to significant categories that emerged from the consultants. Thirteen categories of place association are common to all five cultural groups. Subsequent to these thirteen are categories that are specific to the individual cultural group. The number of the category appears in the approximate location of the cultural resource, and the specific cultural resource is listed in the cultural resources discussion section. This mapping method is designed to facilitate comparison of the cultural resources across the various cultural groups.

Cultural Resources Map Key
Hispanic Community

1. General park knowledge
2. a) Culturally meaningful sites and structures cited by laypersons
   b) Culturally meaningful sites and structures cited by experts
3. Favorite places
4. Nice/beautiful places
5. Places disliked
6. Childhood memories
7. Memories of places that have changed
8. Places to relax
9. Places to take the children
10. Places to take relatives from out-of-town/visited on school trips
11. Overly touristy places (to be avoided)
12. Places associated with local people's work
13. Aesthetically displeasing places

Specific to Hispanic Community
14. Additional sites marked on maps after further probing
15. Places to meet with friends and acquaintances

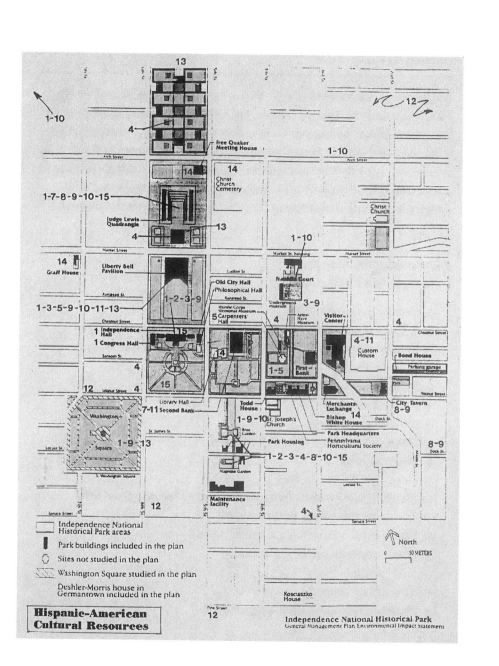

CULTURAL RESOURCES MAP

The following map is organized according to significant categories that emerged from the consultants. Thirteen categories of place association are common to all five cultural groups. Subsequent to these thirteen are categories that are specific to the individual cultural group. The number of the category appears in the approximate location of the cultural resource, and the specific cultural resource is listed in the cultural resources discussion section. This mapping method is designed to facilitate comparison of the cultural resources across the various cultural groups.

Cultural Resources Map Key
Italian American Community

1. General park knowledge
2. a) Culturally meaningful sites and structures cited by laypersons
   b) Culturally meaningful sites and structures cited by experts
3. Favorite places
4. Nice/beautiful places
5. Places disliked
6. Childhood memories
7. Memories of places that have changed
8. Places to relax
9. Places to take the children
10. Places to take relatives from out-of-town/visited on school trips
11. Overly touristy places (to be avoided)
12. Places associated with local people's work
13. Aesthetically displeasing places

Specific to Italian American Community
14. Places/Symbols of patriotic significance

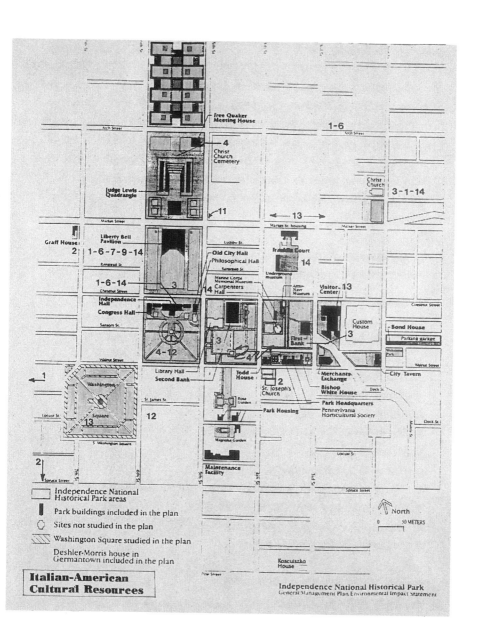

CULTURAL RESOURCES MAP

The following map is organized according to significant categories that emerged from the consultants. Thirteen categories of place association are common to all five cultural groups. Subsequent to these thirteen are categories that are specific to the individual cultural group. The number of the category appears in the approximate location of the cultural resource, and the specific cultural resource is listed in the cultural resources discussion section. This mapping method is designed to facilitate comparison of the cultural resources across the various cultural groups.

Cultural Resources Map Key
Jewish American Community

1. General park knowledge
2. a) Culturally meaningful sites and structures cited by laypersons
   b) Culturally meaningful sites and structures cited by experts
3. Favorite places
4. Nice/beautiful places
5. Places disliked
6. Childhood memories
7. Memories of places that have changed
8. Places to relax
9. Places to take the children
10. Places to take relatives from out-of-town/visited on school trips
11. Overly touristy places (to be avoided)
12. Places associated with local people's work
13. Aesthetically displeasing places

Specific to Jewish American Community
14. Places to walk

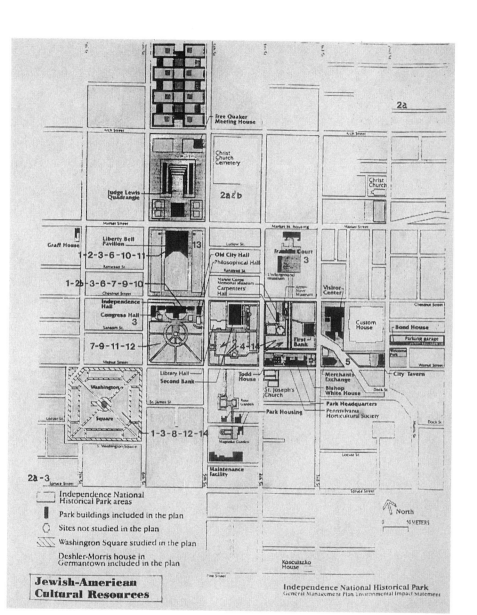

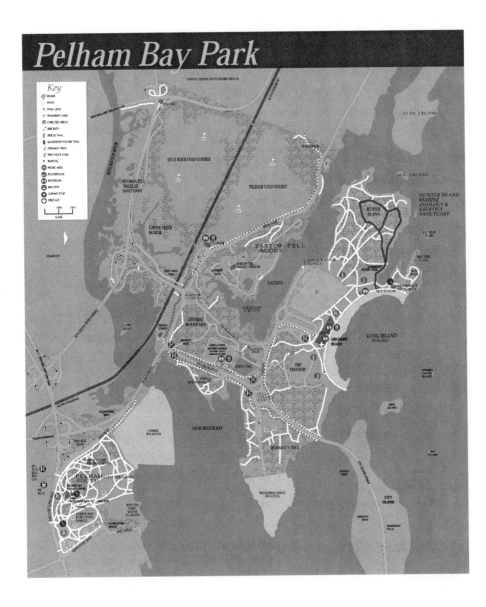

MEANINGS AND SYMBOLS

The Liberty Bell is symbolic for Jewish Americans because its inscription, "Proclaim liberty throughout all the land . . . ," is from the book of Leviticus in the Hebrew scriptures. The transect walker, focus group participants, rabbis, and two of the individuals interviewed all mentioned this. One focus group participant said the Liberty Bell symbolizes "the whole concept of freedom—of not being slaves, the whole idea that motivated the colonists, of going from darkness to light, from exile to redemption—that's part of our religion!"

The transect walker pointed out that the ideas in the Declaration of Independence—especially that all people are created equal—were "the Jewish people's gift to America." She said the Irish historian John Lecky wrote that "Hebraic mortar cemented the foundations of America." She too cited the source of the Liberty Bell's inscription and said, "It seems like it's ringing now, it's so symbolic, even though it hasn't in more than a century. It has more and more meaning around the world." Within the general area, Elfreth's Alley has particular significance for her as a surviving example of the setting in which the pre-Revolutionary Jewish community lived. She notes the homes on Elfreth's Alley are of eighteenth-century Jews, like Jacob Cohen, a fur trader. She thinks of the colonial era as a time of such widespread acceptance of Jews that many gradually assimilated into the Gentile community—for her, the meaning of "Philadelphia" seems to ring true. A symbol of that acceptance, although not a Quaker congregation, is Christ Church (Episcopal), which has an old tradition of an annual dinner with Mikveh Israel congregants; the parish keeps kosher dinnerware for that very purpose. Old City was the home of German Jews at the turn of the twentieth century, she says, while the eastern European immigrants settled in South Philadelphia. Her favorite figure of the Revolutionary period is Franklin, who, among other things, was the largest Gentile contributor toward the retirement of Mikveh Israel's mortgage debt at a time of financial crisis.

A place of obvious importance to Jewish Americans is the NPS-maintained Mikveh Israel Cemetery. When we were there, a woman came in saying she was thrilled to find it unlocked for once: "I consider it an honor to be here. It's one of the prizes of Philadelphia. I always look in here but it's never open."

CULTURAL REPRESENTATION

Rabbi Caine was quite vocal that Mikveh Israel Cemetery should be highlighted and that places related to Chaim Solomon should have the park's signature on them (Solomon was a major raiser of funds for the Revolutionary War). He is particularly sensitive to the Jewish role in the history of the Independence period, and to the absence of its representation in the park. Others consulted were wary of making much of the Jewish contribution. Although many feel proud of it, they fear divisiveness would result from calling attention to specifically Jewish contributions. One man said, "It's a secular place. The importance of what happened there unites us. We want to be treated evenhandedly—if Jewish sites are pointed out, then others should get similar treatment."

Conclusions

The preceding ethnographic accounts give a vivid sense of the diversity of reactions and concerns expressed by consultants within the identified cultural communities, although other communities were excluded from the research process. For instance, two representatives of the Native American community were interviewed at the offices of the United American Indians of Delaware Valley, on the edge of the park. These consultants, of the Lenapi and Nanicoke tribes, were familiar with the park because of the location of the office. They felt the park had alienated Native Americans: for one thing it had done very little to remind the public that the park's land originally belonged to Lenapi Indians. Further, they claimed there is no discussion of Ben Franklin's historical visit to the Iroquois communities, and the ways in which he applied firsthand knowledge of indigenous populations' social organization to political projects.

But even with a strategically "rapid" ethnographic study, we think it is possible to identify the distinct voices of the cultural communities from the quotes and comments, particularly with regard to meanings and symbols and cultural representation. From these findings we have learned that African Americans are the most concerned about their lack of cultural representation in the park's colonial history, that Asian Americans and Hispanic Americans are less directly concerned but would like to see their stories integrated as part of the American experience, while Italian Americans and Jewish Americans are at best ambivalent about presenting themselves as distinct from other Americans. Three of the cultural groups—African Americans, Hispanic Americans, and Jewish Americans—mentioned places they would like to see commemorated or markers they would like to see installed to bring attention to their cultural presence within the park boundaries. And many of the cultural groups—in particular Hispanic Americans, African Americans, and Asian Americans—were anxious to have more programming for children and activities for families. The Hispanic Americans were particularly interested in the recreational potential of the park, and their sentiments were echoed by at least a few consultants in each of the other cultural groups. Overall, there are some distinct messages from each cultural group, as well as some general preferences that relate to the majority of the groups.

Table 7.2 provides an overview of the complex analysis presented in the preceding sections. From these comparisons emerge similarities and differences

Table 7.2. Independence National Historical Park:
Comparison of Cultural Groups by Content Analysis Categories

Category	African Americans	Asian Americans	Hispanic Americans	Italian Americans	Jewish Americans
Visitor Use	Do not visit frequently—too busy, no black history, went in past; infrequent visitation; walk, play in Washington Square	Liberty Bell and Independence Hall main sites visited; visit with guests or on special occasions	Frequent visitation—parade, trysts, work, lunch; bring children to play or out-of-town guest	Most do not visit, a few feel it is not safe; many visited when young; some like scenery	Do not visit frequently; memories of visits; take out-of-towners; some walkers
Meanings and Symbols	History and cultural identity; no particular meaning or negative meaning because only for tourists or whites	Clean, safe, organized, peaceful place; "broken Bell" represents reality of struggle for freedom	Pretty, quiet place; "too serious"; historical place representing struggle for freedom	Some attachment to physical elements; story of immigration, community making, and Constitution	Liberty Bell is symbolic because of inscription; Declaration of Independence very important
Cultural Representation	Does not represent African Americans; feel excluded from park because of lack of diversity	Do not expect to be represented; emphasize education; Chinese feel little recognition of contributions to city	Puerto Rican Day Parade is a kind of representation; would like to see more exhibits about their culture	Balch Institute represents immigrant groups; park should be for everyone; pretzels, pizza, hoagies	Ambivalent about calling attention to group; Mikveh Israel and Chaim Solomon should be highlighted

that were found across the cultural groups. The following generalizations can be drawn from this cross-cultural analysis: 1) most interviewees do not use the park except to take visitors, and although many have fond memories of the past, some now find the park unsafe; 2) most cultural groups feel that the park's meaning is related to the struggle for freedom and relate this history to their own histories; 3) some cultural groups have appropriated the symbol of the Liberty Bell and given it their own cultural meanings; 4) many cultural groups feel excluded from the park because of the lack of cultural representation and identification; and 5) most of the cultural groups would like more participation in the park.

The case study of Independence National Historical Park demonstrates that cultural representation of cultural groups is critical to their use and relationship to the park. The erasure of history documented for the African American and Jewish American communities, and the exclusion of the Hispanic American and Vietnamese American communities through monolingual programs and signage, illuminates how cultural/ethnic groups respond to the cues of the physical and social environment. If we want culturally diverse groups to participate in designed public spaces, then it is the responsibility of the designers and planners, as well as the federal, state, and municipal governments, to take seriously the words of these respondents: Design places that erase our history, and/or create places that exclude us in subtle ways, we will not come. Cultural representation in urban space is material evidence of the history and local politics of exclusion of marginalized and/or minority residents. Urban parks provide social and environmental mnemonics that communicate who should be there, and historical buildings and places, markers, and monuments set the stage for human behavior.

Our research provides clear evidence of how planning and design practices of historic preservation can disrupt a local community's sense of place attachment and disturb expressions of cultural identity for local, ethnic populations. New ethnic and immigrant groups can be excluded because of a lack of sensitivity to cultural barriers such as an inability to read or speak English, nonverbal architectural cues, as well as signs of cultural representation. This ethnographic study of Independence Historical National Park presents an example of the kind of research that can be undertaken to recover histories that have been changed and/or erased during earlier historical preservation or urban renewal projects. We conclude that cultural representation in urban parks is fundamental to their use and maintenance by local groups. Understanding the intimate relationship between ethnic histories, cultural representation, and park use is critical to successful design and planning in any culturally diverse context.

Postscript

One result of this research is that the National Park Service authorized further research on Philadelphia communities and their relationships with the park. As part of this subsequent work, NPS distributed a questionnaire to all the people identified in the REAP study as community consultants. An Independence National Historical Park official said that she often uses the REAP report as a reference to back up arguments made in support of the park's position with regard to ethnic communities. Officially, the REAP report became an appendix to an environmental impact statement on the new General Management Plan. The REAP lent support to the idea of including Washington Square within the national park, a change already proposed at the time that has since gathered momentum. The same NPS official said that the new Liberty Bell enclosure will have a space in it for "public dissent." NPS officials said they felt it was important to include a specific space where dissenting groups could express themselves. One administrator called this an "indirect impact" of the REAP; even though free speech was not an issue the REAP addressed specifically, numerous interviewees "dissented" from the park's patriotic messages.

Note

1. Doris Fanelli at Independence points out, "The story that they were built by slave labor or solely by African Americans is not documented.... The fact that the story is repeated often shows its significance to the group that tells it, and what the park should take from the story is the request for acknowledgment."

Chapter 8
Anthropological Methods for Assessing Cultural Values

Introduction

It is sometimes difficult to find the right method for studying people in a place, especially when you are trying to collect something as sensitive, intangible, and variable as cultural values. The best way to start, however, is to understand what "toolkit" or "palette" of techniques is available, and what works best in diverse fieldwork situations. As researchers, we have had to decide what would work best in a range of settings and have adapted our methods to fit the specific site and problem. Sometimes it was as simple as turning what was to be a focus group into a group interview when surrounded by a group of excited preteens or reworking an interview into a transect walk or bicycle ride with people on the move or exercising in a park. The everyday circumstances of fieldwork make it necessary to be flexible and often creative when faced with problems such as people who do not want to talk to you or are uninterested in your topic. But the benefits outweigh the costs, in that you begin to learn what others think, value, and care about, and with this knowledge you can begin to solve problems and conflicts and empower local communities.

This chapter reviews the available qualitative methods in anthropology for assessing cultural life and values in large urban spaces such as landscape parks, beaches, and national heritage sites. These sites are useful for planning, designing, reconstructing, and managing these complex places. The case studies found in the previous chapters illustrate how these methods were used to answer specific questions and solve problems in specific urban environments, and thus they will provide practical examples of how each method was used in different settings.

We begin with a brief overview of qualitative methods in cultural anthropology. Ethnographic and observational approaches seem most appropriate because they apply to both individual and group levels of analysis. Two other methodologies—constituency analysis and ethnosemantics—are also applicable. The limitations of each is discussed, and a third methodology, the rapid ethnographic assessment procedure, or REAP, is proposed as the most inclusive and useful for solving park problems. REAP methodologies grew out

Figure 8.1. Ethnographers at work at Jacob Riis Park

of agricultural and national park projects, and when applied to planning and design problems integrated elements of constituency analysis used in landscape architecture, and ethnosemantic methodologies used in historic preservation and planning projects. The remainder of the discussion focuses on the REAP.

Anthropological/Ethnographic Methods
Overview of Qualitative Methods in Cultural Anthropology

Qualitative methodologies in cultural anthropology are characterized by their humanism and holism—a philosophical position that argues that humans, and human behavior, cannot be understood or studied outside the context of a person's daily life and activities. Methodological strategies subsumed within this definition include cognitive, observational, phenomenological, historical, ethnographic, and discourse approaches to research. Each of these approaches focuses on distinct aspects of the social world and vary in terms of their appropriateness for particular problems, level of analysis, and available researcher roles. For this discussion, methodologies are arranged in order of their complexity and scope of inquiry beginning with cognitive and observational approaches that focus on one dimension of human activity, a mental or behavioral process, followed by phenomenological and historical approaches that integrate human activity with the environmental context, and concluding with ethnographic and discourse approaches that include human activity, environment, and social, cultural, and/or political context.

Cognitive approaches include both the study of cognition as a mental process, often as reflected in language, and cognition as a set of categories that structure perception through the attribution of meaning. One application is in the area of ethnosemantics—the study of cognitive meaning from the culture's own point of view. Semantics refers to the linguistic analysis of the structure of meaning in a language and culture. Most semantic work is based on intensively interviewing key informants to produce linguistic taxonomies, hierarchies of concepts and terms that describe an individual's understanding of the world, and that collectively describe the culture (Low and Ryan 1985). For instance, a professional working with an anthropologist could develop a taxonomy of house types by asking informants to name all the kinds of houses that exist in their town. Once a list of all the possible house types is developed, the researcher then asks what distinguishes each house type, then repeats the procedure until a complete linguistic map of all housing kinds and their characteristics has been produced. Ethnosemantics refers to a modified semantic procedure that focuses on the semantic structure of one group of people in relation to their local environment. When used in studies of the built environment, it also incorporates the role that language plays not only as a structural or taxonomic system but also as a symbolic communication about important cultural ideas.

Observational methodologies in which overt behavior is observed by the researcher are the mainstay of qualitative research, and include simple observation of activities and behavioral mapping, as well as elaborate systems of time-lapse photography of public spaces (Whyte 1980), ethnoarchaeological techniques (Kent 1984), and nonverbal communication strategies for understanding the built environment (Rapoport 1982; Low 2000). For instance, William H. Whyte (1980) spent seven years filming street behavior with a small movie camera to produce the conclusions presented in *The Social Life of Small Urban Spaces*. The analyses of this film produced a set of urban design principles that have been used as the basis of urban public space zoning in New York City.

Ethnoarchaeological techniques combine traditional archaeological data obtained from on-site excavation and stratification analysis with historical documents and ethnographies of local groups who may be using the site in ways similar to their local ancestors. The idea is to use observations of contemporary people's built environment, everyday behavior, and social and ritual activities to interpret archaeological findings (Kent 1984). Finally, observation of nonverbal behavior has been used to theorize about how people understand a site. Rapoport (1982) argues that fixed features of a site, such as the buildings, trees, and elements that cannot be easily moved, and nonfixed features, such as furniture, produce very different kinds of meanings. Non-fixed features are more important when trying to understand nonverbal communication. In all of these cases, observational techniques are at the core of the research project or theoretical explanation.

Phenomenological approaches differ from observational methods in that the object of study is not separated from the act of perceiving. Studies focus on "place," and "how place grows out of experience, and how in turn, it symbolizes that experience" (Richardson 1984, 65; also see Low 1982). The emphasis is on the individual perceiver and his/her experience as empirical evidence of the world, rather than the observation itself being the evidence separate from the observer. This epistemological difference is quite significant in terms of the way that the researcher records (in field notes and narrative rather than on maps or film) and understands the data collected.

Historical approaches locate a particular site, place, or built form in its temporal context. Historical approaches are very important for architectural historians, archaeologists, and others, because they can provide insight into past values of the sites and how perceptions and significance have changed over time. Historical approaches address the past users and study the material culture and its evolution, but they do not address the current users of the site, who are best understood through ethnographic approaches.

Ethnographic approaches are broader and include the historical as well as the social and political context of the site as a means of understanding contemporary sociocultural patterns and cultural groups. Ethnographic research, the process of describing a culture, has the capacity to accurately predict local responses to design and planning proposals and can help evaluate complex alternatives through systematic cultural understanding. Depending on the magnitude of the geographical area, length of time spent, and historical depth of the study, ethnography produces a complete cultural description of a site, as well as descriptions of interconnected nonlocal communities and relevant adjacent sites. For instance, the ethnographic study of Jacob Riis Park at the edge of Brooklyn and Queens in New York City found that the National Park Service's restoration of Robert Moses's bathhouse was of little importance to the new visitors to the site, who come to the beach to picnic in the shade and to enjoy family activities (figure 8.1). These new users, mostly recent immigrants from Central and South America, are not aware of the history of the site and did not understand the fencing off of the historic Mall area with its direct view of the Empire State Building. Instead, they are upset that so many of the few trees remaining on-site were cordoned off. Their response has been to ignore the fencing and picnic in the trees wherever possible. The ethnographic study illuminated this source of conflict (between those users and park management), thus providing the possibility of better communication, design, and planning of the historic site in the future (see Chapter 5).

Discourse approaches include social experience, the reciprocal acts of speaking and being spoken to, and the emergent product of that speaking, the object of the conversation. They consider the object of study, the text, the context, and interpretation of the object as one continuous domain. Discourse approaches are only beginning to be used in applied settings because of the difficulty of gathering the data and the highly specialized forms of transcription and notation.

In Table 8.1 each methodological approach is evaluated by 1) the focus or scale of the research—individual, group, or societal; 2) the degree of contact and/or involvement with the research "subject"—minimal, moderate, or total; and 3) the kind of problem most often associated with the methodology. The utility of each methodology is derived from the researcher's need to answer questions at a specific scale, in a time frame that controls the degree of involvement, and within the domain of a particular research problem. The application criteria derive from these same decision variables.

These approaches are appropriate for different kinds and levels of research. For instance the individually based methodologies (cognitive, phenomenological, and some discourse methodologies) are excellent for eliciting individual users' experiences and perceptions of the site, while the societally based approaches (historical and discourse approaches) provide methods that uncover issues of cultural significance and social change. All of these methods answer research problems of concern to the planning or design practitioner or manager; however, we would like to highlight the observational and ethnographic approaches that focus on the group and the individual within the group. These two methodologies address the core objective—that is, to identify local site use and disuse and, even more importantly, to understand the motivations, norms, values, intentions, and symbolic meanings underlying that use or disuse. For example, while phenomenological research can elicit statements of place attachment and place identity, ethnographic research describes the place attachment of groups within the geographical community. Further, ethnographic approaches focus on sociocultural values as a central part of the research endeavor.

Ethnography combined with observational methodologies, however, requires considerable time in the field to complete, usually up to a year or more. Working with design and planning professionals—as well as conservation practitioners, park managers, and other professionals—requires brief, direct procedures for understanding a particular site. Two of these strategies have been used in historic landscape preservation projects and are discussed because

Table 8.1. Qualitative Methodologies
in Cultural Anthropology: Research Appropriateness

Methodological Approach	Scale/Level of Inquiry	Degree of Involvement	Research Problem
Cognitive	Individual	Minimal	Rules, ideals, and perceptions
Observational	Group and individual	Minimal	Behavior, observable actions and activity sites
Phenomenological	Individual	Total	Experience of places and events
Historical	Societal	Minimal	Social and cultural trends, comparison of sites
Ethnographic	Group and individual	Moderate	Cultural motivations, norms, values, intentions, symbols and meanings
Discourse	Individual and societal	Moderate	Underlying meanings of speaking/conversation

of their appropriateness—they combine observation and ethnography—and because they offer methodological shortcuts that allow for short-term application during an ongoing, site-specific project.

Constituency Analysis: A Methodology for Landscape Architecture

The development of an appropriate social science research method for landscape architecture began as a consequence of Setha Low, an anthropologist, working with design faculty and students.[1] They needed a way to organize, collect, and conceptualize social data relevant to design problems. Constituency analysis was an attempt to integrate the complex, recursive process of design with social data. Table 8.2 summarizes the five-stage design process with three social data phases—stages I, II, and V—that necessitate anthropological methods.

The first stage is problem formulation composed of client definition and problem clarification. For any project there are a number of possible clients and user groups including a paying client (often the federal government), specific user groups, communities or neighborhoods on or near the proposed site, and often potential regional or national constituencies who may use the site in the future. Interviews, an analysis of influence processes, and other techniques

ANTHROPOLOGICAL METHODS

Table 8.2. Constituency Analysis

Stage I: Problem Formulation	Stage II: Data Collection	Stage III: Programming	Stage IV: Physical Design	Stage V: Evaluation
Client definition	Constituency identification	Data interpretation	Conceptual design	Measurement of change
Problem clarification	Needs and desires assessment	Data application	Physical framework	Interpretation of meaning
	Constituency conflicts			

are necessary to generate a list of all the clients, or stakeholders, involved in the design.

Once the client and problem are defined, the designer begins to collect data on the perceptions of the residents and future users of the site. This data collection stage takes the form of an identification of constituencies and their perceived needs, desires, and social conflicts. Constituency identification is the enumeration and description of the kinds of people living on or near the project site, that is, their social, cultural, and demographic characteristics. Any number of sampling techniques and methods, from participant observation of local communities to a questionnaire survey of randomly selected residents and users, can be employed to collect such data. Once constituencies are described and categorized into groups, the second task of identifying constituency perceptions, needs, and desires begins. This information, which becomes the basis of later physical design decisions, is more difficult to collect in that direct elicitation techniques are not usually successful. The methods suggested for constituency needs and desires assessment are therefore indirect techniques that attempt to stimulate response and opinion concerning possible land use and physical design scenarios such as expert interviews, mental maps of patterns of site utilization and perceptions, and projective tests. A final step in the data collection procedure includes the identification of constituency conflicts concerning issues that impact the future success of any planned change. Depending on the project, an analysis of constituency conflicts may become part of the programming procedure, especially when the project objective is to resolve conflicting land uses.

The third and final stage before implementation and physical design is the construction of a program, a set of specific objectives for the final design. The program orders and applies the constituency needs and desires to particular design decisions. Finally, an evaluation of the design, based on original project objectives and social criteria, requires some form of measurement of social change. A number of anthropological methodologies have been developed to monitor the social impact of large-scale projects including the REAP discussed below. Social change is often measured by a questionnaire survey of previously defined outcome variables. However, qualitative techniques, such as participant observation and structured interviewing, can be used when the design intervention is at a relatively small scale.

Constituency analysis is an excellent system for integrating constituency identification into the planning and design process. The process of client identification is similar to stakeholder identification, and constituency identification, needs and desires assessment, and the working out of constituency conflicts are applicable to most large urban sites. The drawback, however, is that some sites do not have clear constituencies, or that their constituencies do not match or correlate with cultural values on-site. For these reasons, park anthropologists developed methodologies such as the REAP that are more flexible and utilize a wider set of techniques and methods. Nonetheless, the sequencing of stages and the emphasis on the reiterative nature of design and planning problems are useful in thinking about developing a cultural values assessment process.

Ethnosemantic Methodology: Design and Translation at Historic Sites

Ethnosemantic techniques have been used to translate local values into elements of material culture that could then be preserved. The separation of the perceptions of architectural historians and the public is increased by differences in professional and popular culture. Architects and architecturally trained historians, as well as most design, planning, and heritage professionals, participate in a process of socialization that provides a common language, set of symbols, value structure, and code of rituals and taboos. The public does not share this perceptual system but holds images and preferences that are embedded in their own beliefs, customs, and values. Conflict may arise when these two "cultures" compete for control over land use, building, landscape, and/or preservation decisions. In such a situation, the methodological and conceptual skills of someone trained in ethnosemantics or other anthropological and linguistic methodologies are useful to resolve the "cultural" conflict. When park managers and planners face decisions that they know may be fiercely contested, looking for another way to translate the cultural differences, through a method such as one of those described, may solve the disagreement by finding the middle ground or appropriate language necessary to proceed with the plan, design, or other desired change.

Ethnosemantic methodologies assume that culture is encoded in language that can be elicited through a linguistic, taxonomic analysis. Structured questions organize responses into taxonomic categories to create cultural domains of meanings. These methods have been applied in a modified form to historic preservation of buildings and landscapes. Research on the ethnosemantic structure of Greek village houses uncovered their traditional social status meanings (Pavlides and Hesser 1989) and translated culturally appropriate details of eighteenth-century stone farmhouses in a rural Pennsylvania community into standards for in-fill architectural design (Low and Ryan 1985). Both studies began by determining the range of architectural variation in the local community, investigating the local meanings attributed to the variation, and then verified those meanings employing an ethnosemantic method. Pavlides and Hesser (1989) photographed architectural details of Greek village houses that they suspected were symbolic of a family's social standing based on previous interviewing and house survey. They then presented these photographs back to the community and asked them to tell them what each meant. The responses of the community members were used to insure that the researchers' interpretation of symbolic meaning reflected the community's.

Low and Ryan's (1985) study of historic buildings in Oley, Pennsylvania, was designed to elicit what local residents thought were meaningful characteristics of their stone farmhouses. The project was part of a rural preservation program and utilized a historic buildings survey as a guide to architectural variation in the community. A representative blue-ribbon panel was interviewed as to the degree of "Oleyness" for each of the architectural details found in the survey. The research linked architectural elements with cultural images through the exploration of "Oleyness" as a culturally relevant cognitive domain.

Rapid Ethnographic Assessment Procedures
Rapid Assessment and Applied Ethnographic Research

Rapid assessment methodologies have been adapted for research on parks in the United States from methods pioneered in developing nations. The idea of rapid assessment originated at about the same time in two separate fields of work: one in rural and agricultural development projects, the other in connection with public health programs and epidemiology. Rapid assessment concepts have been adapted to nonethnographic contexts as well, such as conservation biology (Abate 1992). Rapid assessment procedures (RAPs) are widely used in the health field; the term and methodology originated in a manual published in 1981 by Susan Scrimshaw and Elena Hurtado, which was first applied in Guatemala and later field-tested in 15 other countries (Macintyre 1995; Manderson 1997).

The term used in agricultural development is rapid rural appraisal (RRA), which originated in workshops held in Sussex, U.K., in 1978 and 1979 (Manderson and Aaby 1992; Beebe 1995). At that time, development officials devised rapid appraisal methods to gather relevant social information in connection with rural development initiatives operating with limited time and resources. All rapid assessment methodologies belong within the rubric of applied research: as Kumar (1993) points out, the task is not to solve theoretical puzzles or generate theory but to reach more rational decision-making processes in real-life circumstances.

Rapid methods grew out of at least two problems: 1) the need to collect and assimilate social and economic information in rapidly changing contexts and 2) the lack of enough anthropologists working in applied medical and health fields, especially in developing countries. A related problem is the length of time and cost required to train field researchers. In the health field, international agencies have worked to develop effective health-education and disease-control programs in developing countries, as well as accurate program-evaluation systems. Recognizing that health involves a sociocultural context, agencies have sought research methodologies from anthropology that provide highly specific social and cultural information (Manderson and Aaby 1992; Harris, Jerome, and Fawcett 1997).

Secondary reasons for the rise of rapid methods include a service agency "culture" that relies on consultancies rather than employment of a permanent research staff and the realization among agency officials that community insiders have valuable access to settings and possess knowledge that may be helpful to program design (Manderson and Aaby 1992). Rapid assessment methods have been widely used internationally for programs dealing with diarrheal disease, nutrition, primary health care, acute respiratory infection, and epilepsy and have been sponsored by such agencies as the United States Agency for International Development, United Nations University, United Nations International Children's Education Fund, and the World Health Organization (Harris, Jerome, and Fawcett 1997).

Within anthropology, rapid assessment methodologies are historically associated with action anthropology, a value-explicit approach that works to achieve self-determination and to foster the accumulation of power in local communities. Anthropologists such as Steve Schensul saw a need for time-effective research techniques, arguing that theoretical elegance and justification back to the theoretical literature did not serve community goals. Schensul devised what he called "commando anthropology" in Chicago in 1973. In that instance, 11 separate research teams entered Chicago public schools all at eleven o'clock to evaluate the English as a second language programs then operating. The findings were used to support a suit filed with the Illinois Civil Rights Commission (Van Willigen 1993).

Like action anthropology, rapid assessment methods place considerable importance on including local people as part of the research team. The premises, preferences, and interests of the powerful parties in a situation—for example, the investigators, governments, and donor agencies—determine the ways these parties construct reality and choose their actions. It therefore becomes important to involve all the different stakeholders in a given situation so as to offset the biases of the funders and investigators. Equally valuable is the discovery of indigenous knowledge held by small farmers, women, and the landless, among others (Schensul 1985; Kumar 1993). The anthropologist is involved in the action at hand but as an auxiliary to local community leadership, using his or her research skills to support the attainment of community goals (Van Willigen 1993).

Rapid assessments differ from traditional qualitative research in that more than one researcher is always involved in an often-multidisciplinary team, research team interaction is critical to the methodology, and the results are produced much faster (Beebe 1995, 42). The two basic methodological principles in rapid research are triangulation of techniques and iteration. Triangulation, or the use of multiple methods, "aims at maximizing the validity and reliability of data" (Manderson 1997, 6). The semistructured interview, expert interview, and community focus group are the characteristic elements of a triangulated methodology. Iteration refers to the constant reevaluation of findings as new data come in, with the implication that new research questions may be generated in light of such reevaluations (Harris, Jerome, and Fawcett 1997; Manderson 1997).

Critics of rapid methods focus on questions of external validity and reliability. Because the research participants are selected on a cluster basis or on other nonprobability criteria, the results are generally considered to be invalid for the total population (Kumar 1993; Manderson 1997). Rapid-methods data give a relatively accurate picture of the prevalence of a phenomenon, attitude, perception, or behavior pattern but not its extent or pervasiveness (Kumar 1993). "Rapid assessments [choose] timely, focused, and qualitative information at the expense of 'scientific' sureness of results through strong probability sampling" (Manderson 1997, 2).

Thus, rapid methods are also held to pose problems for internal, or construct, validity—that is, giving variables or behaviors the right names and assigning accurate meaning to observations (Harris, Jerome, and Fawcett 1997).

In traditional ethnography, the years spent observing and living with the research subjects tend to assure high construct validity, but rapid research can lead to misunderstandings about the phenomena observed. However, using triangulation techniques reduces this risk. Reliability—the ability to produce the same results repeatedly—is also at issue with rapid methods, where the difficulty is attributed to observer bias. The multidisciplinary nature of rapid research teams helps to reduce observer bias (Harris, Jerome, and Fawcett 1997).

In North America rapid methods have been applied to social impact assessments (in the United States, pursuant to the National Environmental Policy Act [NEPA]) and to community needs assessments (Crespi 1987; Liebow 1987). NEPA requires federal agencies to involve the public in decision-making processes. For park managers, conducting ethnographic research in relation to planning and programming decisions complies with NEPA and provides cultural information useful to operating, protecting, and conserving cultural resources (Mitchell 1987).

Ervin (1997) reported on a community needs assessment for Saskatoon, Saskatchewan, which took about six months in all, using a combination of six qualitative methods. The four-person research team worked under contract to the local United Way to rank the community's social service priorities. The resulting report ranked priorities such as eliminating hunger and strengthening preventative services and avoided making direct evaluations of social service providers. Still, some of the stakeholders in the project, dependent on United Way funding, were wary of cooperating with the research effort—and in some cases were hostile to it.

Within the National Park Service, "cultural resource management" (CRM) is concerned with identifying the impact of federal and other development on archaeological sites, historic buildings, and the like and then managing the impact in various ways, as required by federal law (Van Willigen 1993, 164). Cultural anthropologists working in CRM have more recently been applying ethnographic research to contemporary communities and adapting rapid assessment methods as one of several approaches to applied research. The Park Service's Applied Ethnography Program defines seven ethnographic research methodologies, among them the rapid ethnographic assessment procedures (NPS 2000). Each methodology is employed in one way or another to investigate and describe cultural relationships between particular local communities and park resources, sometimes to support nominations of lands and sites to the National Register of Historic Places (Joseph 1997). The REAP is appropriate for project-driven applications because it provides a great deal of cultural information useful to planning purposes within a short time—generally, a four-month time frame (Liebow 1987; NPS 2000). The short time frame of a REAP is a crucial advantage in the event of substantial proposed construction, which involves major commitments of funds, negotiation of political support, and agency commitments as to feasible and timely project development.

The *Cultural Resource Management Bulletin*, an NPS publication, devoted an issue in 1987 to ethnographic research within the agency on contemporary communities. Although REAP in particular is not addressed, several of the articles in the issue elucidate the Park Service's views on the utility of applied ethnography in general. Bean and Vane (1987) and Low (1987) corroborate Van Willigen's (1993) observation that funding for cultural research within the NPS is devoted primarily to historical and archaeological concerns, rather than to the cultural relationships of present-day communities to park resources.

Howell's (1987) report on her experience in 1979 with the Big South Fork National Recreation Area in Tennessee is an example of how important understanding these relationships can be. In that project, researchers were able to convince a cooperating federal agency, the Army Corps of Engineers, to divert a small portion ($50,000) of the project's cultural resources budget to ethnographic research, in the form of a folklife survey. Howell observes that history and archaeology have long had important roles in cultural resource management and interpretation, but until recently little research was done to understand the lifeways of people living in or near national parks. Marlowe and Boyd (1987) allude to the rival "cultures" within NPS. We suppose, too, that park managers tend to see the lifeways of ordinary people as self-evident and to pride themselves on "knowing their people."

NPS first undertook ethnographic research in connection with Native American communities having long-standing associations with certain park lands. These lands and associated cultural resources are required by Native Americans or other local communities for their continued cultural identity and survival. NPS labels these lands "ethnographic resources," and the peoples associated with them "traditionally-associated" or "park-associated" peoples (Crespi 1987). In providing systematic data on local lifeways, applied ethnographic research is intended to enhance the relationships between park management and local communities whose histories and associations with park cultural resources are unknown or poorly understood (Bean and Vane 1987; Crespi 1987; Joseph 1997).

The NPS literature points to several kinds of benefits from ethnographic research. One is in the area of conflict management: for example, when the local community opposed a new park designation, ethnographic knowledge helped management identify opportunities for compromise and potential mitigating measures (Wolf 1987). Another type of benefit involves community empowerment. Joseph (1997) stresses the collaborative nature of the applied ethnographic research done by the Park Service, where ordinary citizens and community leaders participate alongside elected officials, park managers, and the researchers. While the powerful constituencies in a community make their views known, ethnographic research is a vehicle for identifying less visible groups and drawing them into the decision-making process.

A third important benefit of ethnographic research is in the area of finding ways to both present and represent the cultural heritage of local communities within a park's interpretive program. Independence National Historical Park has recently made efforts to represent the history of Philadelphia's African American community in its interpretive program. Such efforts have been controversial among park staff, some of whom feel that the stories of less famous people should not compete with the official focus on the founding of the nation (Blacoe, Toogood, and Brown 1997). Minuteman National Historical Park, in Massachusetts, has restored and preserved farming as a traditional cultural practice within the historic environment the park preserves and interprets. Information that may be uncovered only through ethnography, such as the gendered division of labor on family farms, becomes an important part of the park's interpretive message and is also helpful to effective management (Joseph 1997).

Manderson and Aaby (1992) point to an absence of health-related rapid assessment procedure studies in the literature. As they see it, rapid assessments are done in support of program requirements, not as scientific research, and their frequent use in contracts and consultancies makes them not the usual stuff of academic reporting. Applications of the REAP are also not widely reported. In this book we illuminate the uses and achievements of one REAP in a variety of park studies.

The REAP Methodology

In a REAP a number of methods are selected to produce different types of data from diverse sources that can be triangulated to provide a comprehensive analysis of the site. A description of each method is briefly presented, followed by Table 8.3, which summarizes the product and outcome of each.

HISTORICAL AND ARCHIVAL DOCUMENTS

The collection of historical documents and review of relevant archives, newspapers, and magazines begins the REAP process. At historically significant sites this process may be quite extensive, especially if secondary sources do not exist. The importance of careful historical documentation should be emphasized, since it is through a thorough understanding of the history of the site that areas of cooperation and conflict often become clear and identifiable.

PHYSICAL TRACES MAPPING

Physical traces maps record the presence of liquor bottles, needles, trash, clothing, erosion of plantings, and other traces of activities. These maps are completed based on data collected early in the morning at each site. Records of physical evidence of human activity and presence provide indirect clues as to what goes on at these sites during the night. Physical traces mapping presumes that there is a base map of resources and basic features available that can be used to locate the physical traces. Otherwise, part of the task is to create such a map, both for the physical traces and for the behavioral maps. At many archaeological sites a base map might not be available, adding another step to the research process.

BEHAVIORAL MAPPING

Behavioral maps record people and their activities located in time and space. Such maps arrange data in a way that permits planning and design analyses of the site, and they are very useful in developing familiarity with the everyday activities and problems of a site. They are most effectively used in limited park areas with a variety of social and economic uses where the researcher can return repeatedly to the various social spaces during the day.

TRANSECT WALKS

A transect walk is a record of what a community consultant describes and comments on during a guided walk of the site. The idea is to include one or two community members as research team members, in order to learn about the site from the community member's point of view. In most REAPs local consultants work with the researcher as collaborators. In the transect walk, however, this relationship is particularly important in that the method is dependent on the quality of the relationship between the collaborator and the researcher, and on the ability of the community member to articulate community concerns.

INDIVIDUAL INTERVIEWS

Individual interviews are collected from the identified populations. The sampling strategy, interview schedule, and number of interviews varies from site to site. In most cases on-site users and residents who live near the site are interviewed, but in specific situations interviews might be collected more broadly.

EXPERT INTERVIEWS

Expert interviews are collected from those people identified as having special expertise to comment on the area and its residents and users, such as the head of the vendors' association, neighborhood association presidents, the head of the planning board, teachers in local schools, pastors/ministers of local churches, principals of local schools, and representatives from local parks and institutions.

IMPROMPTU GROUP INTERVIEWS

Impromptu group interviews occur where people gather outside of public places or at special meetings set up with church or school groups. The goal of group interviews (as opposed to individual interviews or focus groups) is to collect data in a group context as well as to provide an educational opportunity for the community. Impromptu group interviews are open-ended and experimental and include any community members who are interested in joining the discussion group.

FOCUS GROUPS

Focus groups are set up with those people who are important in terms of understanding the park site and local population. As opposed to the large, open group interviews, a focus group consists of 6 to 10 individuals selected to represent especially vulnerable populations such as schoolchildren, seniors' groups, and physically challenged groups. The discussions are conducted in the primary language of the group and directed by a facilitator and are usually tape-recorded.

PARTICIPANT OBSERVATION

The researchers maintain field journals that record their observations and impressions of everyday life in the park. They also keep records of their experiences as they interact with users and communities. Participant observation is a valuable adjunct to the behavioral maps and interviews. It provides contextual information and data that can be compared to what is seen and enables accurate data interpretation.

ANALYSIS

Interview data are organized by coding all responses and then content-analyzed by cultural/ethnic group and study question. Transect walks, tours, and interviews are used to produce cultural resource maps for each group. Focus groups determine the extent of cultural knowledge in the community and identify the areas of conflict and disagreement within the community. Mapping, transect walks, individual and expert interviews, and focus groups provide independent bodies of data that can be compared and contrasted, thus improving the validity and reliability of data collected from a relatively small sample. As in all ethnographic research, the use of interviews, observations, and field notes, as well as knowledge of cultural group patterns and local politics, are used to help interpret the data collected.

A number of procedures are used to analyze the data. First, the resource maps are produced by an overlay method that combines the behavioral maps, physical traces maps, and participant observation notes. These maps are descriptive in that they summarize activities and disruptions on-site. Second, a research meeting is held in which participants summarize what they have found in their interviews. These are general observations that guide the research team (or researcher) as they begin to develop more precise coding strategies. This synthetic stage is quite important in that it provides a place to start thinking about what has been found. These "general summaries" are used to explore theoretical approaches and prioritize the coding procedure.

The third step is to take each generalization and break it into a set of codes that can be used to analyze the field notes. Once this is completed, the interview questions are then reviewed and a similar coding scheme is developed. The interview coding relies both on the findings of the maps and field notes and on the structure of the questions themselves. This is the lengthy part of the analysis process, and it requires discussion of the research team with the client and, in some cases, individual stakeholders. Some coding schemes may require multidimensional scaling and a quantitative analysis, although qualitative content analysis is usually adequate in a REAP. Because the REAP is a "rapid" procedure, the number of interviews is usually under 150 and therefore can be analyzed by hand. The advantage of a qualitative analysis procedure is that the data are not abstracted from their context and therefore retain their validity and detail. The final step is a triangulation of the various analyses and a search for common elements, patterns of behaviors, and the identification of areas of conflict and differences, both in the nature of the data and among the groups themselves.

The question of who should be undertaking these various projects does not have a simple answer. The overall project including the identification of stakeholders, the development of a values typology, the values assessment process, the evaluation and ranking of values, and a follow-up with more detailed assessment as necessary should be organized and directed by the professional. But values assessment, particularly when using a REAP, is a team process. Experienced ethnographers and field-workers will be able to produce the necessary data more quickly and easily. Further, the analysis process requires considerable training and background in qualitative analysis techniques. On the other hand, the techniques involved in constituency analysis, ethnosemantic methodologies, and REAPs can be learned through a series of training workshops. Local participants can become excellent on-site field-workers,

Table 8.3. Overview of Methods, Data, Products, and What Can Be Learned

Method	Data	Product	What Can Be Learned
Historical Documents	Newspaper clippings, collection of books and articles, reading notes	History of the site's relationship to the surrounding communities	Historical context for current study and planning process
Physical Traces Mapping	Collected trash, patterns of erosion	Description of nighttime activities on-site	Evening activities not observed
Behavioral Mapping	Time/space maps of sites	Description of daily activities on-site	Cultural activities on-site
Transect Walks	Transcribed interviews and consultant's map of site	Description of site from community member's point of view	Community-centered understanding of the site; local meaning
Individual Interviews	Interview sheets	Description of responses of the cultural groups	Community responses and interest in the park
Expert Interviews	In-depth interview transcriptions	Description of responses of local institutions and community leaders	Community leaders' interest in park planning process
Impromptu Group Interviews	Transcription of meeting	Description of group perspective; educational value	Group consensus of issues and problems
Focus Groups	Tape-recorded and transcribed	Description of issues that emerge in small-group discussion	Elicits conflicts and disagreement within the cultural group
Participant Observation	Field notes	Sociocultural description of the context	Provides context for study and identifies community concerns

and the REAP process usually includes local collaborators. In fact, part of the point of undertaking a REAP is to create connections to the local community. The best situation, if finances allow, is to bring together a team made up of professional(s), ethnographer(s) (number would depend on language demands), and two or three local residents and/or experts who would like to be part of the values assessment process. The residents and experts can be trained by the ethnographers to assist in interviewing and mapping, while the ethnographer would undertake the group interviews, focus groups, and participant observation. There are many useful combinations of expertise, and each has to be developed on-site to fit the circumstances, as the case studies in this volume have demonstrated.

Note

1. Adapted from Low 1982.

Chapter 9
Conclusion
Lessons on Culture and Diversity

William H. Whyte's seminal work in the 1970s on small urban spaces was so clear and convincing that the city of New York revised its zoning code to reflect most of his recommendations. Whyte's work inspired some of his associates to found the Project for Public Spaces, a consulting firm that has worked to bring his vision of user-friendly, comfortable, and popular public spaces to communities throughout metropolitan New York and beyond. With this book we at the Public Space Research Group seek to expand the dialogue about public spaces beyond the issues of comfort and vitality propounded by Whyte, as important as they are.

We realize that it is one thing to talk about comfort and vitality in the public realm and quite another to discuss race, ethnicity, class, and exclusion. However difficult, these issues are becoming increasingly pressing as private groups take over from public agencies in planning, designing, and managing large public spaces. The spaces William Whyte studied in New York were privately built and managed—that is, plazas and other spaces provided by private owners of large office buildings for public use. Our concern is with truly public spaces, including the great urban landscape parks built by the city decades ago and always completely public. Private advocacy and management has brought some of these parks back from conditions of neglect and underutilization. Central Park in particular is the restored jewel in the crown of New York's public parks, thanks to the highly successful career of the Central Park Conservancy. But renovation and restoration of a space can affect its cultural equilibrium. Attractive as it is, is Central Park as inclusive and democratic a space now as it was before the advent of the Central Park Conservancy? Some people dislike questions like that—how dare one question a group that has produced such wonderful results in a beloved public space? So what if a few people feel less welcome?—look at how many more people overall come to the park now; so the argument goes.

We do not argue for a return to the run-down and dangerous park conditions of the 1960s and 1970s. We commend the tireless efforts of friends-of-the-park groups and conservancies everywhere in bringing urban parks back from the edge. What we hope to do here is to demonstrate how important it is to maintain the cultural diversity that makes these great spaces truly urban. We think that most park advocates share in this vision of culturally diverse public spaces. We also think that many do not understand how the reconstructions and assertive management techniques can encode symbols of class privilege and so discourage and even exclude many people of color, immigrants, and poor and working-class people, all of whom should be as welcome in the public spaces of the city as the assimilated white professionals who support the park conservancies.

Lessons Revisited

Having concluded our tour of parks and research methodologies, we return to the six lessons for promoting and sustaining cultural diversity in parks and heritage sites, introduced in Chapter 1. In the following discussion we restate each lesson and then elaborate on it, referring to our various examples from field research.

(1.)

If people are not represented in historical national parks and monuments or, more importantly, if their histories are erased, they will not use the park. The classic example of erasure and consequent nonattendance in our work is that of African Americans in Philadelphia in relationship to Independence National Historical Park. In that case the coincidence of several factors had the effect of erasing not only the symbols but the black community itself from the neighborhood of the park. The area just south of what is now the park contained the historic black settlement in Philadelphia, which W. E. B. Du Bois described in the *Philadelphia Negro* in 1897. The Society Hill redevelopment project and the federal action of land acquisition and clearance for the park itself displaced what remained of this African American community in the 1950s. What took their place was a shrine to the white founding fathers of the United States. Independence National Historical Park and the adjacent Society Hill neighborhood together replaced an unvarnished swath of the historical city with a colonial park and upscale neighborhood, peopled largely by white residents and white tourists. Until recent years Independence National Historical Park gave scant attention to the roles of nonwhites or women in interpreting the founding of the nation. The park lacked markers to commemorate important African American contributions, such as the construction of Independence Hall. Visitors could come away from the park without realizing that African Americans even lived in Philadelphia during the era of the Revolution.

LESSONS ON CULTURE AND DIVERSITY

In reviewing the case studies discussed in this book, we see other kinds of erasure as well. In Prospect Park, the restoration of original design elements caused the removal of authentic parts of the park's material fabric for their failure to be sufficiently historic—unwanted buildings and facilities, 1930s park benches, and soon the 1950s skating rink. Erasure may also include the failure to recognize or interpret the social uses of a place. The National Park Service operates Jacob Riis Park with little attention to its historical importance to long-established patron groups such as the gay community, African Americans, or Irish American and Italian American youth from southern Brooklyn. The park's Robert Moses pedigree is a priority that limits management's ability to maintain and adapt the park to current needs, yet even that history is not marked or interpreted. There are no signs, markers, or exhibits at Riis Park about Moses, or Riis himself, or about the vibrant gay bays and the black bays of the 1970s. Millions of dollars were spent to partially restore the Moses-period bathhouse, but still there is no interpretive signage. Many visitors would appreciate the chance to learn more about the history of the bathhouse and the beach.

(II.)

Access is as much about economics and cultural patterns of park use as circulation and transportation; thus, income and visitation patterns must be taken into consideration when providing access for all social groups. At Ellis Island, poor residents living a mile away in Jersey City are seldom seen among the middle-class visitors who have typically traveled hundreds, even thousands of miles to get there. There are various reasons why people of color may be underrepresented at a park or heritage site, one of which is the question of representation discussed above. Economics are a separate but equally important factor in nonattendance. Underrepresentation is sometimes attributed to cultural patterns when economics may be the real cause. Most parks have no admission fees per se, even national parks with costly facilities, programs, and large rosters of staff (although many charge a sizable vehicle entrance fee). Ellis Island is as free as Independence, but visitors must pay a steep ferry fare to get there. That, we discovered, is a significant barrier for poor city residents. Cultural patterns are important as well—as at Independence, people of color might be less interested in the story of European immigration told at Ellis Island and the Statue of Liberty. Even so, people would have visited Ellis Island at least for the views, cool breezes, and access to the water were it not so expensive to get there.

Circulation and transportation sometimes *are* the main problems: inadequate public transportation has been considered the biggest obstacle in the way of Gateway National Recreation Area reaching its potential as a national park for city residents. Public transportation to much of Gateway is slow, especially on weekends when visitation is highest. At Ellis Island transportation is not the problem: the ferry is easily accessed from both Manhattan and Jersey City. In this case, cost is the factor that sifts the poor out of the visitor population.

(III.)

The social interaction of diverse groups can be maintained and enhanced by providing safe, spatially adequate territories for everyone within the larger space of the overall site. Park managers tend not to think in these terms, concentrating instead on the needs of the resource, that is, the material thing itself. We have watched over the years as management in Prospect Park tries to displace users from certain territories because of a mismatch between vernacular activity and the prescribed use of the area. Because the historic design included dense perimeter plantings, management has limited picnickers in the Vanderbilt entrance area by letting the forest reclaim much formerly open ground and by leaving new park benches largely out of its facility renovations. Management has also curtailed the volleyball playing by Mexican immigrants near 15th Street by roping off areas of lawn and planting trees where the games are played; this because the historic design calls for scenic landscape "passages." Yet the demand for social uses of the park is heavy enough to overcome these frustrations, and people form their territories anyway. What if, on the other hand, management decided to *encourage* these uses? Management could provide places to play volleyball, perhaps a high-impact surface, where playing volleyball would not result in a ruined lawn. At Riis Park, the Park Service has now expanded picnicking facilities in the back-beach areas as a result of our REAP research in 2000, providing many more tables and grills, even shelters. At the same time, the public desire to picnic in the shade of trees continues to conflict with the historic preservation imperative of protecting scenic landscape passages from intrusive use. Thus, the wooded margins along the Mall remain fenced to keep picnickers out.

(IV.)

Accommodating the differences in the ways social classes and ethnic groups use and value public sites is essential to making decisions that sustain cultural and social diversity. Sustaining diversity in parks can be an important part of sustaining diversity in the city overall. In Pelham Bay Park the users of Orchard Beach are largely Latino, and we suppose that many whites do not go there because they perceive it as a Latino beach. There are some whites among the user population, however, some of whom maintain a sense that the park is for them by coming at nonpeak times and by patronizing out-of-the-way areas like Hunter Island.

The park management has been sensitive to the needs of such groups, as in the example related in Chapter 6 where white community seniors have been allowed to use the Ecology Center as a sort of clubhouse. The most striking aspect of the history of Pelham Bay Park is the contrast between large-scale government planning and construction, on the one hand, and the roles played by different user groups in establishing local territories in the park—not just through recurring occupancy but by modifying the space to suit their purposes. This history includes campers maintaining tent colonies and gardeners undertaking their own landscape architecture. Parks that operate under the burden of historic designs, such as Jacob Riis Park and its better-funded New York cousins, Prospect Park and Central Park, tend to make design integrity a priority, sometimes at the expense of user preferences. Both Pelham Bay and Van Cortlandt parks lack this design legacy, and management there has more freedom to let the users appropriate park space.

People need to feel that a public park is for them. As privatized managements redesign parks in a polite, upper-class idiom, poorer people and people of color may read the landscape as exclusive—something for others. The contrast of Battery Park and the Battery Park City parks in Manhattan offers a case in point. Battery Park, one of New York's oldest, has been a completely public park, maintained by the city and used by a great variety of publics. Next to it along the Hudson River shoreline are the parks of Battery Park City, built on reclaimed land in the past 20 years. The Battery Park City parks were designed by prominent landscape architects and are maintained to an exceptionally high standard thanks to revenues generated by Battery Park City real estate. Clearly, the Battery Park City parks exist to enhance the appeal of Battery Park City for its overwhelmingly white and Asian American, affluent residents and office tenants. During a research project in Battery Park and Battery Park City in 2002, we observed a pronounced difference in user class and ethnicity between the two parks (Low, Taplin, and Lamb 2005). Battery Park City is not gated; there is no one to tell certain people not to enter. Still, people read cues of exclusivity in the landscape. The working-class users of Battery Park, predominantly black and Latino, stay there for the most part, despite the proximity and relative comfort of the adjacent parks in Battery Park City. They leave Battery Park City parks largely to Manhattan's affluent professional class.

(v.)

Contemporary historic preservation should not concentrate on restoring the scenic features without also restoring the facilities and diversions that attract people to a park. In Pelham Bay Park we found no conflict between recreational values and those of historic preservation, as the targets of recent restorations have been the recreational facilities at Orchard Beach. Making these facilities more functional serves the social and recreational needs of the park users. Prospect Park tells a different story—there restoration has been practiced with primary allegiance to the park's design legacy as a work of art. The management has restored a variety of landscapes in the park, including the elaborate turn-of-the-century entrance structures and attendant sidewalks in some of the perimeter areas. Although intended mainly for aesthetic effect rather than to promote social activity, these perimeter restorations do make the park more attractive to all its users.

The largest and costliest piece of restoration, however, involves the picturesque water features and woodland landscapes at the center of the park. Restoration in these areas conflicts with sociability in at least two ways: 1), by restricting people from the restored areas for years at a time and then allowing a very limited range of movement along fenced paths and 2), by not including places of congregation and assembly within the restorations. Management has been selective in its restorations, prioritizing water features, woodland hillsides, and rustic bridges. Work on the many places designed to attract and accommodate people has been left for another day. Olmsted and Vaux provided attractions and comfortable gathering places for numbers of visitors within and throughout their pastoral and picturesque landscapes in Prospect Park. We argue that restoration that includes social values as well as aesthetic and ecological values can be just as true to the original design and meet historic preservation criteria as well.

(vi.)

Symbolic ways of communicating cultural meaning are an important dimension of place attachment that can be fostered to promote cultural diversity. The symbols themselves are typically balloons, banners, or similar visual devices put up by visiting parties for the duration of their visit, after which they remove them. As shown in Chapter 6, these symbolic communications are significant in establishing the cultural scene; they communicate a shared identity to friends and, in many cases, to strangers. They may also communicate a welcome to strangers of different identities.

Visitors often attach their banners to trees, an act that usually passes without reaction from park officials. In Prospect and Pelham Bay parks, the staff fear the potential damage to trees from people climbing on them, swinging from their branches, and lighting cooking fires next to their trunks. Tying banners or birthday party balloons to branches is a lesser concern. At Riis Park, where the management has reached a state of alarm over the visitors' use of trees, especially those uses that cause the trees greater stresses: attaching heavy tarps for shade or attaching hammocks to the branches. The black pines planted in the sands of Riis Park are small trees relative to the big hardwoods in Pelham Bay and Prospect parks; also, they have been dying from a blight. Thus the trees at Riis Park are withering from an unusual set of stresses, both ecological and behavioral.

Permanent cultural symbols are another matter altogether. They are the stock-in-trade of heritage parks like Independence and Ellis Island, and the park management maintains control over the symbolic discourse. Questions of symbolic representation of ethnic groups and women, for example, can become pressing issues, as they have at Independence, but no one other than the Park Service erects or displays permanent symbols and markers. As we have seen, there is more latitude in recreational parks, especially ones like the Bronx parks that have never been subject to a totalizing design. Surely the long-lasting German tent colonies of the early twentieth century in Pelham Bay Park used symbolic expressions—the tents themselves, perhaps banners—to communicate messages about identity and territoriality.

Prospect Park, although it bears a famous and jealously defended design, has two interesting examples of more or less permanent, symbolic cultural expressions of and by users. These are the drummers' circle and the Haitian roots music circle, both located near the lakeshore on the east side of the park. Both sites were originally created by park users from nearby communities. The drummers' circle communicates pan-African inclusiveness and a welcome to all. The Haitian circle appears to be a local cultural symbol; its focal element, the Gran Bwa, was understood only by those within the group.[1]

Prospect Park has allowed user groups to make these material changes to the park space. We urge the park management to take further steps in this direction, sharing with user groups like these the prerogative of interpreting the culture of the park and inscribing that interpretation in the landscape.

Ethnography in National and Local Parks

Public space ethnography is a valuable tool in a variety of settings, from local recreational parks to national heritage parks. The National Park Service has an ethnography program to study relationships between park resources and user communities. The logic of doing so became clear once anthropologists realized that Native American communities in the western United States had ongoing cultural relationships with park resources—in many cases, sacred sites. Realizing that, it would be important to understand such relationships before undertaking some plan or program that would in any way disrupt the material fabric of such locations.

It is more difficult to discern such cultural relationships in complex urban settings such as those discussed in this volume. The park resources are often constructed, changeable, and utilitarian, rather than pristine and awesome natural landscapes. Similarly, the cultural identities of heterogeneous, multicultural urban communities would not seem to depend especially on particular locations amid the constant flux of urban life. People move and neighborhoods change; the industries and commercial institutions that seemed so permanent in one generation may have disappeared in the next.

Still, place matters. People remain social beings forming communities *in place,* and the more places change, the more people seek to preserve familiar, everyday landscapes. There is constant conflict in urban places between the market-driven forces of change and people's efforts to keep things as they are. Private property is increasingly affected by the public urge to maintain stability through legal mechanisms such as zoning, historic preservation, and environmental impact reviews and through protest, opposition, and "nimbyism."[2] Decisions about public property are even more contentious—witness the breadth of public discussion over prominent sites such as replacing the World Trade Center, memorials on the Washington Mall, and where to move the Liberty Bell. The urge to preserve public landscapes like Prospect Park takes the form of reconstructing an idealized version of its original landscape architecture. Thus attachments to public places are various and strong. We can begin to understand the depth and dimensions of place attachment by investigating a variety of public places.

The national and municipal parks surveyed in this volume differ in their character and in the relationship their managers have to ethnographic research. Although overshadowed by the Park Service's concern with archaeology, the applied ethnography program is well established and active. Municipal park agencies in New York City and elsewhere lack such an established applied ethnography investigative arm. As discussed earlier in this book, some city parks—those with their own administrators—conduct user surveys from time to time that solicit attitudes toward park services. The literature indicates that in at least some other U.S. cities, notably Chicago and elsewhere in the Midwest, there is a tradition of sociological user research in municipal parks.

Different Park Types

Parks differ in character and purpose. While municipal parks mainly provide recreation, national parks enshrine places important to the national identity. Yellowstone and other natural parks preserve symbolic landscapes. Many patriotic and historical themes are encoded in the great western parks: discovery and exploration, conquest, the frontier and westward expansion, nature and wilderness values, national grandeur, rugged individualism, and so on. Although they deal with built environments, national heritage parks like Independence National Historical Park and the Statue of Liberty and Ellis Island National Monument have similar missions of preserving national symbols and educating the public about historical events. Calling these places "parks" is almost a misnomer; they are educational heritage sites that do little to provide for leisure recreation in the tradition of a park.

Gateway National Recreation Area represents a third category, a kind of hybrid of national and municipal park, that preserves resources of national significance but also provides for leisure recreation, just as a municipal park does. Other examples of the type are the Golden Gate National Recreation Area and the Boston Harbor Islands National Park Area. These parks bring the resources of the National Park System to urban populations who, it is thought, would not otherwise have national park experiences. As physical spaces they are very different from traditional municipal or national parks. Rather than reserving a contiguous space solely as a park, these are noncontiguous collections of separate properties, including surplus military installations, nature reserves, and sites formerly operated by local park agencies. Interspersed among them, at least at Gateway, are residential communities such as the Breezy Point Cooperative at the western end of the Rockaway barrier island. In effect, people are living within the park, and adjacent park lands can become neighborhood territory.

Park Layout

This more ambiguous pattern, of discontinuous park space interspersed with established communities, is increasingly characteristic of new park development. By contrast, the classical example of park development in New York is that of Central Park, where the city took title to a single piece of territory, evicted the occupants, razed all urban land uses within its borders, and then built the park. Prospect Park was developed in the same way. Today evicting residents and businesses, to say nothing of whole communities, to create a park is politically and ethically impossible. Instead, parks are established around existing communities. At the Fire Island National Seashore, the National Park Service took title to certain then-undeveloped lands on Fire Island to keep them as open space. At Cape Cod National Seashore many preexisting private residences remain within park territory, but no other land can be developed for private residential purposes. In the Delaware Water Gap National Recreation Area, the National Park Service owns very little territory, instead providing a planning and administrative overlay intended to provide for a variety of recreational opportunities and to protect the characteristic historic landscape of the area from inappropriate new development.

Such arrangements have been used for areas of unusual scenic value, as at Fire Island and Cape Cod, and for areas having historic landscapes and historic cultural resources, such as the Delaware Water Gap NRA. Similarly, Minuteman National Historical Park, in Massachusetts, preserves sites and landscapes associated with the American Revolution, and the Blackstone Valley National Historic Corridor, in Massachusetts and Rhode Island, preserves historic features of a river valley associated with the industrial revolution in New England. At Minuteman, the Park Service owns and manages the core historic sites but cooperates with local governments and private landowners in planning for the larger area. At Blackstone Valley, the Park Service role is largely one of planning, oversight, public education, and administration, rather than landownership and management.

Pelham Bay and Van Cortlandt parks were both developed by the city of New York in the late nineteenth century, basically following the model set by Central Park, but without the compelling logic of reserving such relatively large territories solely for park use. As a result, neither park has been able to function as a single, integral park space. Van Cortlandt, especially, has lost its intended spatial integrity to the exigencies of road building and water supply development. Had these parks been developed in recent times, we suspect they would have followed one of the newer models of discontinuous park land coinciding with preexisting residences and other urban uses.

Politics of Funding and Service to Park Users

National parks operate in a political context very different from that of municipal parks. National parks have federal funding and are much less dependent on or responsive to local political conditions. Most municipal parks depend on local public funds; even a park like Prospect Park, whose privatized management raises lots of money for the park, taps mainly public sources of funding. If parks are important to local constituencies, moneys for reconstruction and operation will be found. The New York City Department of Parks and Recreation, for example, has found capital funds to reconstruct much of the recreational infrastructure at Orchard Beach in Pelham Bay Park.

Jacob Riis Park and Orchard Beach were both built by Robert Moses during the Great Depression. In the 1960s and 1970s, as these public beaches lost their original white middle-class constituencies, they also lost funding and became run down. Including Riis Park within the new Gateway National Recreation Area seemed to be the solution for the problems at Riis Park, but as explained elsewhere in this volume (Chapter 2, Chapter 5), including Riis Park within the National Park System has not yet succeeded in regenerating this urban beach.

The contrast between Riis Park and Orchard Beach also illustrates how different management structures respond to changing demographics among users. As a national park unit, Riis Park reflects the strengths of the National Park System in its planning processes. In ethnographic research alone, Gateway commissioned a needs assessment for Riis and other park units in 1995, followed in 2000 by the REAP we conducted for Riis Park. NPS also has rigorous yet innovative criteria for protecting historic cultural resources. However, the Park Service professionals at Riis Park have had difficulty adapting the park to suit current vernacular uses. Part of the problem is attributable to the historic landscape preservation mandate, and part to the rigidities of the NPS management structure with the comprehensive policies, regulations, and procedures under which it operates.

Municipal parks, for the most part, show more flexibility in responding to changing user needs. Management at Pelham Bay Park, for example, sees the cultural expressions of Latino park users as part of the park's identity, and so as something to encourage and support. Park management here can be pragmatic and flexible in making Orchard Beach responsive to local cultural vernacular.

Ethnographic Perspectives and the Value of Ethnography for Park Studies

Understanding the concept of culture is central to understanding the importance of park ethnography, but culture is a complex idea that is often misunderstood. For some, culture is conceptualized as high art and life in a civilized society. Within this paradigm, it is thought that some have more culture than others; minority and often marginalized populations in the city tend to be viewed as deficient in their possession of culture defined in this way.

An alternative view is to think of culture as an abstract package of values, practices, and lifeways that are employed to survive within a particular environment. Understanding these local beliefs and practices enables one to describe and analyze the everyday life experiences of cultural groups as distinct from one's own. From this analysis it is possible to appreciate why some cultural groups use public space in one way and why others use it in another. The concept of culture defined in this second mode provides clues to the presence and dynamics of cultural diversity in large urban public spaces.

Park ethnography is a methodology that focuses on the role of culture in diverse groups' perspectives and approaches toward using park resources. Ethnographic research methods such as participant observation and unstructured interviewing uncover values and behaviors of visitor groups not necessarily captured by other techniques. By engaging in conversation in natural, everyday situations it is possible to discover the categories, ranking systems, cultural frameworks, and systems of meaning that are important to local cultural groups and individual users. For example, by conducting in-depth interviews and group discussions in Vietnamese with residents living near Independence National Historical Park, new cultural meanings attached to the Liberty Bell emerged. We learned that the Vietnamese Americans view it as a symbol of flawed craftsmanship and cultural arrogance, yet at the same time they also saw it as a symbol of colonial independence and freedom and related it to their own colonial struggles.

Researchers discover cultural groups' systems of meaning by sharing experiences with visitors through participant observation. Participant observation, the key method for conducting ethnographic research, captures what can be considered local or alternative forms of knowledge, different from that obtained from questionnaires or structured interviews. And when combined with open-ended interviews, ethnographic research enables researchers to understand the extent to which visitors' verbal expressions and actual behaviors correspond. For example, during interviews in the back-beach areas at Jacob Riis Park, Latino immigrant visitors often limited their criticisms of park management and suggestions for improvement. They usually praised the park and its resources. However, while barbecuing many broke park rules by using garbage bins as barbecue pits, tossing coals in the grass, and throwing cardboard into the trees as a way to create shade. Rangers passing by would tell visitors to put out their fires and to dismantle their informal awnings. But minutes after rangers left the area, picnickers would restart their fires and reconstruct the artificial shade. Clearly, these visitors' needs and desires were at odds with park rules and managers. And there was a discontinuity between their desire to go on record with positive views of the park and their resistance to following park rules and regulations. Perhaps the difference between what they said and what they did reflects their distrust of the park administration, and a belief that their suggestions would not be taken seriously. This disjuncture would not have been uncovered without participant observation. Thus, ethnographic research at parks, beaches, and heritage sites presents a more complete picture of visitor groups and their cultural values and behaviors that have an impact.

Ethnographic research also reflects the people and voices that are not often seen or heard. Active community members know how to participate, but newcomers to the city often do not, or are not certain that their concerns will be recognized. When information is relayed within the context of an ethnographic picture of park users and their everyday experiences, managers can begin to imagine others' experiences and can improve their relationship to their constituents. In the case of Independence, managers were surprised to learn that African Americans and Asian Americans felt excluded from the park. One manager said, "Hearing visitors' quotes was like having spears thrown into my heart." Many managers are deeply and personally invested in their work. The positive side to this is that they care about their work and go to great lengths to make changes that enhance visitors' experiences. Unfortunately this personal investment can also bias their understanding of what is happening in a park, and leave managers psychologically vulnerable to visitors' criticisms of their heartfelt work. Ethnographic research can help managers understand potential biases and/or confirm their anecdotal evidence and perceptions of the park with more systematically collected evidence. It can help managers view and digest information that they might not have been aware of before the beginning of the project.

Finally, ethnography fosters an ethos of caring and consideration between park users and managers. Conducting ethnographic research requires making human connections. Good ethnography entails constituents and researchers engaging with each other as authentic people. Researchers tend to care deeply about people's stories and experiences, and park visitors often appreciate the opportunity to be heard and the possibility to make a difference. With such exchanges ongoing in the park, park users and managers deepen their awareness of their impact on the park. Ethnography helps to make the park a place where people and managers care about their environment together.

The Importance of Cultural Diversity

In reviewing the theoretical rationale for promoting and sustaining cultural diversity discussed in Chapter 1, we found that all of the theoretical and ethical positions have been drawn upon. A number of researchers agree with our use of "social sustainability" and have gone even further in their analyses of the importance of an ecological perspective on park development. Galen Cranz and Michael Boland (2003) have even added a fifth park type to Cranz's original stages of park evolution to include the "ecological park," in which managers are concerned with utilizing renewable resources, self-sufficiency, creating a sustainable ecosystem, and learning about the natural environment. Although Cranz and Boland use the term *sustainability* to refer to the natural systems of park maintenance and continuity and see parks as part of the entire urban ecosystem, they agree that social sustainability is part of this model because of the importance of involving diverse people and cultural groups. Maybe there should be a sixth park model, the "culturally diverse park," where management is concerned with cultural and social sustainability, community participation, and users' needs and desires. Even better, the culturally diverse and ecological models might merge as we refine our understanding of contemporary park development. The notion of what cultural diversity is "good for" (Hannerz 1996) could also be expanded by Cranz and Boland's ecological model in which biological diversity plays an important role. Both of the city parks discussed in this volume, Prospect and Pelham Bay, could be classified as representative of the proposed "ecological" or "socially sustainable" park type based on their struggles to maintain self-sufficiency, promote natural ecological processes, accommodate all users, and encourage visitation from diverse groups of people.

Based on our work in Prospect Park, Jacob Riis Park, and Orchard Beach, we also found that community participation and empowerment within the parks are basic components of creating citizenship and political entitlement. Regardless of whether it is Latino immigrants struggling with park management for more picnic space and tables in Jacob Riis, or old-timers who have adopted the Ecology Center as their own in Pelham Bay Park, we found that these engagements with specific cared-about places produced more community involvement and a stronger sense of national and local identity. Robin Bachin argues that historically parks cemented new relationships as "shared civic space allowed different ethnic groups to make use of local parks to express their heritage and traditions, but it also offered a place in which ethnic difference could be overcome" (2003, 16). It is clear that even today urban parks are places where emerging citizens learn about coexistence, cooperation, and tolerance through park activism and participation.

The issue of cultural property rights and ethical concerns about whose history should be interpreted in a park landscape emerged mainly in the national parks such as Independence and Ellis Island. The erasure of historic artifacts and the sacred burying ground of African Americans in downtown Philadelphia to create the colonial Independence park is one example where cultural property rights discourse could have been useful to work out what would be an appropriate remedy for the future. But cultural property rights arguments were not employed directly by park managers or by community interviewees.

On the other hand, African American and Jewish American interviewees at Independence did draw upon notions of dissonant heritage and the politics of meaning when voicing their feelings of exclusion from the park and its interpretation. African Americans wondered whether their history as slaves who worked on the plantations of the signers of the Declaration of Independence, their lack of citizenship at a moment of emancipation for the white colonists, and the role of free blacks in the building of Carpenters' Hall was a history that some (white) Americans wanted to forget. The disinheritance that occurred at Independence, the conscious ignoring of a shameful history, can be explained by the theoretical work of Tunbridge and Ashworth (1996) and Kenneth Foote (1997) discussed in Chapter 1.

Finally, we found that the cultural values literature offered us the most practical theory and methodology for arguing for the importance of cultural diversity. Values have been mostly discussed in relation to historic preservation and heritage parks by Randall Mason (2002) and Marta de la Torre (2002), but their contributions lie at the heart of our research and analysis. Setha Low was an early participant in these discussions and has presented much of this work to her colleagues over the past ten years. And the cases of the Ellis Island Bridge Proposal, Prospect Park, and to some degree Pelham Bay Park rely heavily on cultural values as an explanatory framework for cultural activities and preferences.

Parks and Democracy

This book is about protecting and sustaining an urban public realm that attracts, supports, and expresses cultural diversity. In our own city of New York, we find that too many spaces in the city center no longer fulfill these goals. Between the movement to shift responsibility for maintenance to private groups and the increased surveillance and other security measures in the post-9/11 era, the cultural and social diversity of New York's public spaces has dwindled. The larger parks, mostly outside the city center, remain hospitable to diverse groups of users—for now. We have tried to show how the various processes of inclusion and exclusion work in a variety of settings. In bringing these processes to light we hope to heighten the awareness of how parks function in bringing people together.

This is an old idea. The great landscape architect Frederick Law Olmsted spoke and wrote extensively on the purposes of parks. Parks would make the city healthier and exert a soothing influence on the weary and wary city dweller. A well-designed park system would provide a framework for urban development of high quality. One of Olmsted's major purposes, however, was to provide a meeting ground where the diverse citizens of a democratic society could come together. Olmsted believed that a complex web of volunteer and recreational social activity, and the communication such activity fostered, was the crucial underpinning of a democratic society. Libraries, reading groups, gymnasiums, game clubs, boat clubs, ball clubs, and so on were all examples of communicative associations—what today is often called social capital. Olmsted believed that parks were fertile social spaces where many such associations could take place. He thought it important to have relaxed, expansive, green settings where people of different backgrounds could encounter each other without the wariness and suspicion that arises in congested urban environments (Gopnick 1997; Olmsted 1997).

In their social history of Central Park, Rosenzweig and Blackmar (1992) agree that providing democratic spaces for "gregarious recreation" was Olmsted's ultimate goal. Central Park is "one of the great interracial and interclass meeting grounds of New York" (Rosenzweig and Blackmar 1992, 475) because of its ability to bring together so many different kinds of people in an amicable setting.

The issues are not so different today, when the population is again swollen by immigration. But the centrifugally expanding cities of today lack the spatial cohesiveness of the nineteenth-century city: people are scattered over far-flung metropolitan patchworks and linked together by highways and telecommunication lines rather than streets, squares, churches, and taverns. Even more than in Olmsted's day, large parks and beaches are so important for their ability to bring together diverse groups where, as Olmsted argued, they can encounter each other in an open and inviting atmosphere. Cultural diversity is a new term but it expresses the old idea that, at the grassroots level, democracy consists of groups of people engaging with one another to make community. Parks and beaches of the types investigated here are vital settings for the fundamental social activity of a democratic society.

Notes

1. The Gran Bwa was a tree stump symbolically carved by a local folk artist. It was at the center of a log circle used throughout the 1990s by Haitian musicians. The stump is now so rotted that the carving is entirely lost.

2. The expression *nimby*, for "*not in my back yard*," refers to a generalized sentiment of resistance to new development of all kinds.

References Cited

Abate, Tom. 1992. Environmental Rapid-Assessment Programs Have Appeal and Critics: Are They the Domain of the Conservation Elite? *BioScience* 42:486–489.

Altman, Irwin, and Setha M. Low. 1992. *Place Attachment*. New York: Plenum Press.

Bachin, Robin F. 2003. Cultivating Unity: The Changing Role of Parks in Urban America. *Places* 15 (3): 12–17.

Barlett, Peggy, and Geoffrey W. Chase. 2004. *Strategies for Sustainability: Stories from the Ivory Tower*. Cambridge: MIT Press.

Bean, Lowell J., and Sylvia B. Vane. 1987. Ethnography and the NPS: Opportunities and Obligations. *CRM Bulletin* 10 (1): 34–36.

Beebe, James. 1995. Basic Concepts and Techniques of Rapid Appraisal. *Human Organization* 54:42–51.

Bennett, John W. 1968. Reciprocal Economic Exchanges among North American Agricultural Operators. *Southwestern Journal of Anthropology* 24:276–309.

Blacoe, Joanne, Anna C. Toogood, and Sharon Brown. 1997. African-American History at Independence NHP. *CRM Bulletin* 20 (2): 45–47.

Blockson, C. L. 1992. *Philadelphia's Guide: African-American State Historical Markers*. Philadelphia: Pearl Pressman Liberty.

Borrini-Feyerabend, Grazia. ed. 1997. *Beyond Fences: Seeking Social Sustainability in Conservation*. Gland, Switzerland: IUCN.

Brill, Michael. 1989. Transformation, Nostalgia, and Illusion in Public Life and Public Space. In *Public Spaces and Places*, ed. Irwin Altman and E. Zube, 7–29. New York: Plenum Press.

Caro, Robert. 1974. *The Power Broker: Robert Moses and the Fall of New York*. New York: Vintage Books.

Carrier, James. 1993. Rituals of Christmas Giving. In *Unwrapping Christmas*, ed. Miller and Daniel, 55–74. Oxford: Clarendon Press.

Cheek, N., and W. R. Burch. 1976. *The Social Organization of Leisure in Human Society*. New York: Harper and Row.

Cohen, Yehudi. 1968. *Man in Adaptation*. Chicago: Aldine.

Cranz, Galen. 1982. *The Politics of Park Design: A History of Urban Parks in America*. Cambridge: MIT Press.

Cranz, Galen, and Michael Boland. 2003. The Ecological Park as an Emerging Type. *Places* 15 (3): 44–47.

Crespi, M. 1987. Ethnography and the NPS: A Growing Partnership. *CRM Bulletin* 10 (1): 1–4.

Cruikshank, Ken, and Nancy Bouchier. 2001. The Heritage of the People Closed against Them: Class, Environment, and the Shaping of a Summer Playground for Hamilton, the Burlington Beach, 1870s–1980s. *Urban History Review* 30:40–55.

Cushing, Elizabeth Hope. 1988. "So Near the Metropolis": Lynn Woods, a Sylvan Gem in an Urban Setting. *Arnoldia* 48 (4): 37–51.

Cutler, P. 1985. *The Public Landscape of the New Deal*. New Haven: Yale University Press.

de la Torre, Marta, ed. 2002. *Assessing the Values of Cultural Heritage*. Los Angeles: The Getty Conservation Institute.

Domosh, M. 1996. *Invented Cities: The Creation of Landscape in Nineteenth-Century New York and Boston*. New Haven: Yale University Press.

Edgerton, R. 1979. *Alone Together: Social Order on an Urban Beach*. Berkeley: University of California Press.

Eliot, C. W. 1999. *Charles Eliot: Landscape Architect*. Amherst: University of Massachusetts Press.

Ervin, Alexander M. 1997. Trying the Impossible: Relatively "Rapid" Methods in a City-wide Needs Assessment. *Human Organization* 56:379–387.

Feather, N. T. 1992. Values, Valences, Expectations, and Actions. *Journal of Social Issues* 48:109–124.

Floyd, M., K. J. Shinew, F. McGuire, and F. Noe. 1994. Race, Class, and Leisure Preferences: Marginality and Ethnicity Revisited. *Journal of Leisure Research* 26, 158–173.

Foote, Kenneth E. 1997. *Shadowed Ground: America's Landscapes of Violence and Tragedy*. Austin: University of Texas Press.

Foresta, R. 1984. *America's National Parks and Their Keepers*. Washington, DC: Resources for the Future, Inc., Johns Hopkins University Press.

Gantt, Harvey. 1993. Reassessing Our Agenda. *Preservation Forum* 7 (1): 6–11.

Gobster, P., and A. Delgado. 1993. Ethnicity and Recreation Use in Chicago's Lincoln Park: In-Park User Survey Findings. *Managing Urban and High-use Recreation Settings*. North Central Forest Experiment Station, USDA Forest Service.

Gopnick, A. 1997. A Critic at Large: Olmsted's Trip: How Did a News Reporter Come to Create Central Park? *New Yorker*, March 31, pp. 96–104.

REFERENCES CITED

Graff, M. M. 1985. *Central Park, Prospect Park: A New Perspective.* New York: Greensward Foundation.

Graham, B., G. J. Ashworth, and J. E. Tunbridge. 2000. *A Geography of Heritage: Power, Culture, and Economy.* London: Arnold Publications.

Greiff, Constance M. 1987. *Independence: The Creation of a National Park.* Philadelphia: University of Pennsylvania Press.

Haglund, K. 1993. *Inventing the Charles River.* Cambridge: MIT Press.

Hannerz, Ulf. 1996. *Transnational Connections: Culture, People, Places.* London: Routledge.

Harris, Kari J., Norge W. Jerome, and Stephen B. Fawcett. 1997. Rapid Assessment Procedures: A Review and Critique. *Human Organization* 56: 375–378.

Harrison, Barbara. 2001. *Collaborative Programs in Indigenous Communities: From Fieldwork to Practice.* Walnut Creek, CA: Altamira Press.

Hayden, Dolores. 1990. Using Ethnic History to Understand Urban Landscapes. *Places* 7:11–37.

———. 1995. *The Power of Place.* Cambridge: MIT Press.

Hayward, Jeff. 1989. Urban Parks: Research, Planning, and Social Change. In *Public Spaces and Places,* ed. I. Altman and E. Zube, 193–216. New York: Plenum Press.

Heckscher, August. 1977. *Open Spaces: The Life of American Cities.* New York: Harper and Row.

Holleran, Michael. 1998. *Boston's "Changeful Times": The Origins of Preservation and Planning in America.* Baltimore: Johns Hopkins University Press.

Howell, Benita J. 1987. Folklife in Planning. *CRM Bulletin* 10 (1): 20–22.

Hutchison, R. 1987. Ethnicity and Urban Recreation: Whites, Blacks, and Hispanics in Chicago's Public Parks. *Journal of Leisure Research* 19 (3): 205–222.

Huxtable, Ada L. 1997. *The Unreal America.* New York: The New Press.

Jackson, J. B. 1984. *Discovering the Vernacular Landscape.* New Haven: Yale University Press.

———. 1997. *Landscape in Sight: Looking at America.* New Haven: Yale University Press.

Johnston, Chris, and Kristal Buckley. 2001. Communities: Parochial, Passionate, Committed, and Ignored. *Historical Environment* 15 (1–2): 88–96.

Johnston, Chris, and Annie Clarke. 2001. *Taking Action: Involving People in Local Heritage Places.* Canberra: Australian National University.

Joseph, Rebecca. 1997. Cranberry Bogs to Parks: Ethnography and Women's History. *CRM Bulletin* 20:20–24.

Karp, Ivan. 1992. Introduction: Museums and Communities: The Politics of

Public Culture. In *Museums and Communities: The Politics of Public Culture*, ed. Ivan Karp, Christine M. Kreamer, and Steven D. Lavine, 1–17. Washington: Smithsonian Institution Press.

Kent, Susan. 1984. *Analyzing Activity Areas: An Ethnoarchaeological Study of the Use of Space*. Albuquerque, NM: University of New Mexico Press.

King County Landmarks and Heritage Program. 1999. *Community Cultural Planning for Heritage Organizations*. Seattle: King County Office of Cultural Resources.

Kornblum, William. 1975. Special Study. *The 1974 Summer Season: Gateway National Recreation Area, New York–New Jersey*. Denver: National Park Service.

Kumar, Krishna. 1993. *Rapid Appraisal Methods*. Washington: World Bank.

Landscape Architecture. 1998. The Park Process: Master Plan for Forest Park, St. Louis. *Landscape Architecture*, January 1998, 26, 28–31.

Lane, Frenchman, and Associates, Inc. 1992. *Cultural Landscape Report, Jacob Riis Park*. Washington: Denver Service Center, National Park Service, U.S. Department of Interior.

Lawrence, Denise, and Setha M. Low. 1990. The Built Environment and Spatial Form. *Annual Review of Anthropology* 19:453–505.

Lefkowitz, Joel. 2003. *Ethics and Values in Industrial-Organizational Psychology*. Mahwah, NJ: Lawrence Erlbaum Associates.

Liebow, E. 1987. Social Impact Assessment. *CRM Bulletin* 10 (1): 23–26.

Loukaitou-Sideris, Anastasia. 1995. Urban Form and Social Context: Cultural Differentiation in the Uses of Urban Parks. *Journal of Planning Education and Research* 14:89–102.

Loukaitou-Sideris, Anastasia, and Gail Dansbury. 1995–1996. Lost Streets of Bunker Hill. *California History* 74 (4): 394–407, 448.

Low, Setha. 1981. Anthropology as a New Technology in Landscape Planning. In *Proceedings of the Regional Section of the American Society of Landscape Architecture*, ed. J. Fabos, 125–134. Washington, DC: American Society of Landscape Architecture.

Low, Setha. 1982. Social Science Methods in Landscape Architecture Design. *Landscape Planning* 31:37–48.

———. 1987. A Cultural Landscapes Mandate for Action. *CRM Bulletin* 10 (1): 30–33.

———. 1992. Symbolic Ties that Bind. In *Place Attachment*, ed. Irwin Altman and Setha M. Low, 165–185. New York: Plenum Press.

———. 1994. Cultural Conservation of Place. In *Conserving Culture: A New Discourse on Heritage*, ed. Mary Hufford, 66–77. Chicago: University of Illinois Press.

———. 2000. *On the Plaza: The Politics of Public Space and Culture.* Austin: University of Texas Press.

Low, Setha M., and I. Altman. 1992. Place Attachment: A Conceptual Inquiry. In *Place Attachment,* ed. Irwin Altman and Setha M. Low, 1–12. New York: Plenum Press.

Low, Setha M., and W. Ryan. 1985. Noticing without Looking: A Methodology for the Integration of Architectural and Local Perceptions in Oley, Pennsylvania. *Journal of Architectural Planning and Research* 2:23–22.

Low, Setha M., Dana Taplin, and Mike Lamb. 2005. Battery Park City: A Rapid Ethnographic Assessment of the Community Impact of 9/11. *Urban Affairs Review:* 655–682.

Lubar, Harvey. 1986. Building Orchard Beach. *Bronx County Historical Society Journal* 23 (2): 75–83.

Macintyre, Kate. 1995. The Case for Rapid Assessment Survey for Family Planning Program Evaluation. Annual Meeting of the Population Association of America.

Mackintosh, Barry. 1991. *The National Parks: Shaping the System.* Washington: National Park Service, U.S. Department of the Interior.

Manderson, Lenore. 1997. *Population and Reproductive Health Programmes: Applying Rapid Anthropological Assessment Procedures.* New York: United Nations Population Fund Technical Report.

Manderson, Lenore, and Peter Aaby. 1992. An Epidemic in the Field? Rapid Assessment Procedures and Health Research. *Social Science and Medicine* 35:839–850.

Marlowe, Gertrude W., and Kim Q. Boyd. 1987. Maggie L. Walker. *CRM Bulletin* 10 (1): 9–11.

Mason, Randall. 2002. Assessing Values in Conservation Planning: Methodological Issues and Choices. In *Assessing the Values of Cultural Heritage,* ed. Marta de la Torre, 5–30. Los Angeles: The Getty Conservation Institute.

Mitchell, Joan. 1987. Planning at Canyon de Chelly National Monument. *CRM Bulletin* 10 (1): 40.

National Park Service. 1994. NPS-28. *Cultural Resource Management Guideline: Applied Ethnography Program.* Washington: U.S. Department of the Interior.

National Park Service. 1995. *Draft General Management Plan, Environmental Impact Statement: Independence National Historical Park, Philadelphia.* Washington: U.S. Department of the Interior.

National Park Service. 2000. *Applied Ethnography Program.* Washington: U.S. Department of the Interior.

Netting, Robert M. 1993. *Smallholders, Householders: Farm Families and the*

Ecology of Intensive, Sustainable Agriculture. Stanford: Stanford University Press.

Newton, N. 1971. *Design on the Land: The Development of Landscape Architecture.* Cambridge, MA: Belknap Press.

New York City Parks and Recreation Department. 1986. *Pelham Bay Park: History.* New York.

Olmsted, F. L. 1997. Public Parks and the Enlargement of Towns, February 25, 1870. In *The Papers of Frederick Law Olmsted,* supplementary series, ed. C. Beveridge and C. Hoffman, Baltimore: Johns Hopkins University Press.

Pavlides, E., and J. E. Hesser. 1989. Vernacular Architecture as an Expression of Its Social Context in Eressos, Greece. In *Housing, Culture, and Design: A Comparative Perspective,* ed. Setha M. Low and E. Chambers, 357–374. Philadelphia: University of Pennsylvania Press.

Phelts, Marsha D. 1997. *An American Beach for African Americans.* Gainesville: University Press of Florida.

Pierre-Pierre, Garry. 1993. A Neighborhood Changes, but Deep-Rooted Residents Remain. *New York Times,* May 30.

Proshansky, H. M., A. K. Fabian, and R. Kaminoff. 1983. Place-Identity: Physical World Socialization of the Self. *Journal of Environmental Psychology* 3:57–83.

Prospect Park Alliance. 1995. *Annual Report.* Brooklyn, NY.

Rapoport, A. 1982. *The Meaning of the Built Environment.* Beverly Hills: Sage Publications.

Richardson, M. 1984. Place, Experience, and Symbol. *Geoscience and Man* 241 (3): 63–67.

Rosenzweig, R., and E. Blackmar. 1992. *The Park and the People: A History of Central Park.* New York: Henry Holt and Co.

Rymer, Russ. 1998. *American Beach: A Saga of Race, Wealth, and Memory.* New York: HarperCollins.

Sarkissian, Wendy, and Donald Perlgut. 1986. *Community Participation in Practice: Handbook.* 2nd ed. St. Kilda, Australia: Impact Press.

Schensul, Stephen L. 1985. Science, Theory, and Application in Anthropology. *American Behavioral Scientist* 29:164–185.

Schnitz, Ann, and Robert Loeb. 1984. More Public Parks! The First New York Environmental Movement. *Bronx County Historical Society Journal* 21 (2): 51–63.

Scott, Catherine. 1993. A Salute to Rodman's Neck. *Bronx County Historical Society Journal* 30 (2): 51–60.

———. 1999. *Images of America: City Island and Orchard Beach.* Charleston, SC: Arcadia.

Sims, James W. 1986. Tent City at Orchard Beach. *Bronx County Historical Society Journal* 23 (1): 5–7.

Smith, Neil. 1996. *The New Urban Frontier*. London: Routledge.

Tate, Alan. 2001. *Great City Parks*. London and New York: Spon Press.

Taylor, D. 1993. Urban Park Use: Race, Ancestry, and Gender. *Managing Urban and High-use Recreation Settings*. North Central Forest Experiment Station, USDA Forest Service.

———. 1999. Central Park as a Model for Social Control: Urban Parks, Social Class, and Leisure Behavior in Nineteenth-Century America. *Journal of Leisure Research* 31 (4): 420–477.

———. 2000. Meeting the Challenge of Wild Land Recreation Management: Demographic Shifts and Social Inequality. *Journal of Leisure Research* 32 (1): 171–179.

Terrie, P. 1994. *Forever Wild: A Cultural History of the Wilderness in the Adirondacks*. Syracuse: Syracuse University Press.

Throsby, David. 1995. Culture, Economics, and Sustainability. *Journal of Cultural Economics* 19:199–206.

———. 1999a. Cultural Capital. *Journal of Cultural Economics* 23:3–12.

———. 1999b. Cultural Capital and Sustainability Concepts in the Economics of Cultural Heritage. 1–36. Paper prepared for the Economics of Cultural Heritage Project, the Getty Conservation Institute.

Tunbridge, J. E., and G. J. Ashworth. 1996. *Dissonant Heritage: The Management of the Past as a Resource in Conflict*. Chichester, NY: John Wiley and Sons.

Van Willigen, John. 1993. *Applied Anthropology: An Introduction*. Westport, CT: Bergin and Garvey.

Von Hoffman, A. 1994. *Local Attachments: The Making of an Urban Neighborhood, 1850 to 1920*. Baltimore: Johns Hopkins University Press.

Warner, Sam Bass. 1968. *The Private City: Philadelphia in Three Periods of Its Growth*. Philadelphia: University of Pennsylvania Press.

———. 1993. Public Park Inventions: Past and Future. In *The Once and Future Park*, ed. D. Karasove and S. Waryan. Minneapolis: Walker Art Center.

Warren, Karen. 1989. A Philosophical Perspective on the Ethics and Resolution of Cultural Properties. In *The Ethics of Collecting Cultural Property*, ed. Phyllis M. Messenger, 1–26. Albuquerque: University of New Mexico Press.

Washburne, Randel F. 1978. Black Under-Participation in Wild Land Recreation: Alternative Explanations. *Leisure Sciences* 1 (2): 178–189.

West, P. 1989. Urban Region Parks and Black Minorities: Subculture, Marginality, and Interracial Relations in Park Use in the Detroit Metropolitan Area. *Leisure Sciences* 11:11–28.

Whitaker, B., and K. Browne. 1971. *Parks Are for People*. New York: Schocken Books.
Whyte, W. H. 1980. *The Social Life of Small Urban Spaces*. Washington, DC: Conservation Foundation.
Wilson, A. 1992. *The Culture of Nature: North American Landscape from Disney to the* Exxon Valdez. Cambridge, MA: Blackwell.
Wolf, Janet C. 1987. Martin Luther King, Jr. *CRM Bulletin* 10 (1): 12–13.
Woolf, J. 1996. In Defense of the Metropolitan Mosaic. *National Parks* 70 (Jan.–Feb.): 41.
Wrenn, Tony P. 1975. *General History: The Jamaica Bay, Breezy Point, and Staten Island Units, Gateway National Recreation Area, New York*. Washington: National Park Service, U.S. Department of the Interior.
Yuval-Davis, Nira. 1998. Diversity, Positioning, and Citizenship. In *Cultural Diversity and Citizenship,* ed. Susan Wright, 22–28. Birmingham, U.K.: University of Birmingham.
Zaitzevsky, C. 1982. *Frederick Law Olmsted and the Boston Park System*. Cambridge, MA: Belknap Press.
Zukin, Sharon. 1991. *Landscapes of Power*. Berkeley: University of California Press.

Index

action anthropology, 184
activity-based groups
  Battery Park constituency groups, 77–82
  fishing, 144
  Liberty State Park constituency groups, 84–91
  naturalists, Orchard Beach, 127, 144–145
  picnicking, Jacob Riis Park, 106, 117–121, 123–125, 206
  at Prospect Park, 39, 59–60
African Americans. *See* cultural groups
Albright, Horace, 30
American Indians. *See* cultural groups
Antiquities Act of 1906, 29
Atlanta, Georgia. *See* parks: Chattahoochee River National Recreation Area
Asian Americans. *See* cultural groups

Battery Park. *See* parks
beaches. *See* parks: American Beach; Cape Cod National Seashore; Cape Hatteras National Seashore; Fire Island National Seashore; Glen Island; Gateway National Recreation Area; Indiana Dunes National Lakeshore; Jones Beach; Lake Itasca State Park; Pelham Bay Park, Orchard Beach; Point Reyes National Seashore
Boston, Massachusetts
  Middlesex Fells and Lynn Woods Reservations, 63
  West End outdoor gymnasium, 26
  See also cemeteries; Eliot, Charles; Olmsted, Frederick Law; parks: Arnold Arboretum; Boston Common; Boston Harbor Islands National Park Area; Franklin Park; Metropolitan Park Commission; park types: state parks and reservations
Breezy Point. *See* New York City: Queens
Buffalo, New York. *See* parks: Delaware Park

Caribbeans. *See* cultural groups
Central Americans. *See* cultural groups
Central Park. *See* parks
cemeteries, 20–21
Chicago, Illinois
  applied anthropology in public schools in, 184
  playground movement in, 26
  South Park District, 27
  See also parks: Chicago Forest Preserve; Indiana Dunes National Lakeshore; Lincoln Park
citizenship. *See* democracy
Cleveland, Ohio. *See* parks: Cuyahoga Valley National Recreation Area
cognitive research methods. *See* research methods
commando anthropology. *See* action anthropology
community participation and empowerment, 11–13, 187, 208
  Charleston Principles, 12
  collaborative research, 187–188
  in Independence National Historical Park research, 151, 170–172
conflict management, 187
Costa Rica. *See* parks: Parque Central
cultural diversity
  lessons for, 5, 195–210
  limitations of, 40–43
  uses of, 16–18
cultural ecology, 5–8, 201
  at Jacob Riis Park, 111–119, 124

cultural ecology (*continued*)
  in Prospect Park, 65–66
  urban beaches, 101
cultural groups
  definition of, 150
  African Americans
    confronting racism, 42, 61–62
    cultural activity in parks, 54–56, 67
    geographic distribution in Brooklyn, 60
    at Jacob Riis Park, 109, 112, 114–115, 119–120
    in Jersey City neighborhoods, 92–93
    at Orchard Beach, 127, 131, 144
    park-related values, 52, 64–65
    in Philadelphia, 149–152, 156–160, 170–172, 188, 196, 207, 209
    underparticipation in wild land parks, 41–43
    use of local parks, 41–42, 57
  Asian-Americans
    in Philadelphia, 151, 160–163, 170–172, 206
    residents of Battery Park City, 199
    use of local parks, 41
  Caribbeans
    cultural activity in parks, 54, 56, 61, 67
    at Jacob Riis Park, 114, 119, 120, 123, 127
    in Lafayette neighborhood, Jersey City, 92
    at Orchard Beach, 137
    use of local parks, 43
  Central and South Americans, 120–121, 124, 137, 178, 198
  Eastern Europeans, 120, 123
    Polish residents in Jersey City, 91–92
  gays and lesbians, 109, 111–112, 125, 197
  Haitians, 54, 201
  Hispanics and Latinos
    geographic distribution in Brooklyn, 60
    at Jacob Riis Park, 110, 112, 119–123, 206
    in Lafayette neighborhood, Jersey City, 92
    park-related values at Liberty State Park, 86–87
    park-related values at Prospect Park, 52–54, 63–65
    in Philadelphia, 151–152, 157, 163–165, 170–172
    at Prospect Park, 208
    Puerto Ricans, 120, 136, 150, 152
    researching among, 206–207
    use of local parks, 41, 58
  Indians, 119, 123
  Italian Americans
    at Jacob Riis Park, 109, 197
    at Orchard Beach, 136, 140
    in Philadelphia, 165–167
  Jewish Americans
    exodus from Flatbush, Brooklyn, 60
    in Philadelphia, 149, 151–152, 167–172, 209
    political and social construction of, 15–16
  Native Americans, 131, 170, 187, 202
  Russians
    at Jacob Riis Park, 116, 119, 121, 123
  seniors
    at Orchard Beach, 127, 140–143, 145, 147, 198
  territorialism, 10, 112, 124, 126, 141,
    at Orchard Beach, 198
  Vietnamese Americans
    in Philadelphia, 151–152, 206
  Western Europeans, 136, 140
  white populations, 115, 120, 140, 197–199
    disproportionate use of national parks, 41–43
    fear of black visitors, 32, 62
    geographic distribution in Brooklyn, 60–61
    park activity, 53–54
    park-related values, 52, 64–65
    in Paulus Hook neighborhood, Jersey City, 91
cultural property rights, 208
  cultural property claims, 10
  explanation of, 9–11
Cultural Resource Management (CRM), 186–188
cultural resources
  definition of, 104, 205
  Jacob Riis Park, 103–104
cultural values, 209
  cultural hegemony, 15–16

INDEX

of groups related to Battery Park, 79–82
of groups related to Independence
  National Historical Park, 158–167,
  169–172
of groups related to Jacob Riis Park,
  111–126
of groups related to Liberty State Park,
  97–99
of groups related to Orchard Beach,
  138–147
of groups related to Prospect Park
  50–53
with regard to landscape, 62–64
culture
  definition of, 150, 205
  of service agencies, 184

democracy, 209–210
  citizenship, 12, 102
  parks as training grounds for, 46
  value of parks for, 46
Detroit, Michigan
  park use by black residents, 42
discourse research approaches. *See* research
  methods
dissonant heritage, 13
DuBois, W. E. B., 196

Eastern Europeans. *See* cultural groups
Eliot, Charles, 24
  landscape design philosophy of, 63
Ellis Island. *See* parks
ethnicity
  definition of, 150
ethnographic research methods. *See*
  research methods
ethnographic resources, 187
ethnography,
  definition of, 179
  for parks, 206–207
exclusive landscapes, 199

Floyd Bennett Field. *See* parks: Gateway
  National Recreation Area
Fort Tilden. *See* parks: Gateway National
  Recreation Area
Fort Wagner. *See* parks: Gateway National
  Recreation Area
Freeport, Long Island, 28

Gateway National Recreation Area. *See* parks
gays and lesbians. *See* cultural groups
gentrification
  Brooklyn, 60
  Jersey City, 92
  Philadelphia, 156
Great Society, the, 30
Green-Wood Cemetery. *See* cemeteries

Haitians. *See* cultural groups
heritage parks. *See* park types
Hetch Hetchy Valley, 29
Hispanics. *See* cultural groups
historical research methods. *See* research
  methods
historic landscape preservation
  conflicts with recreational uses, 200
  cultural values, 209
  Ellis Island, 34
  Gateway National Recreation Area, 34–35
  Independence National Historical Park,
    34, 149–150, 155, 172
  Jacob Riis Bathhouse, 105–106, 108, 110,
    115
  Jacob Riis Park, 103–106
  principles of, 200, 205
  Prospect Park, 48, 66–67, 197, 200
  research methods for, 179–180
  resistance to Ellis Island bridge, 16
Historic Sites Act of 1935, 30
Hoboken, New Jersey, 22
humanism and holism
  definition of, 176

immigrant populations. *See* cultural
  groups
Independence National Historical Park.
  See parks
Indians. *See* cultural groups
interpretive programs
  Gateway National Recreation Area, 104
  Independence National Historical Park,
    149, 188
Italian Americans. *See* cultural groups
iteration, 185

Jacob Riis, 102, 197
Jacob Riis Park. *See* parks: Gateway National
  Recreation Area

Jamaica Bay. *See* parks: Gateway National Recreation Area
Jersey City, New Jersey, 197, 198
   neighborhoods, 91–93
   residents' attitudes toward proposed bridge, 93–96
   See also parks: Liberty State Park
Jewish Americans. *See* cultural groups
Johnson, Lyndon B., 30, 69

Lafayette. *See* Jersey City, New Jersey: neighborhoods
landscape architecture
   constituency analysis methodology, 180–182
Latinos. *See* cultural groups (Hispanics and Latinos)
Laurel Hill Cemetery. *See* cemeteries
Liberty State Park. *See* parks
Lila Wallace-Readers Digest Fund, 39
Long Island Sound, 27
Los Angeles, California, 149
   See also Santa Monica Mountains
Louisville, Kentucky. *See* parks: Cherokee, Iroquois, and Shawnee Parks

Mather, Stephen, 25, 30
Mohonk Mountain House, 45–46
Moses, Robert, 205
   career of, 27–28
   construction of recreational parks, 27
   role in developing Jacob Riis Park, 108, 110, 124
   role in developing Orchard Beach, 27, 134
Mount Auburn Cemetery. *See* cemeteries
Muir, John, 29

National Environmental Policy Act of 1969 (NEPA), 186
National Park Service
   Applied Ethnography Program, 186–188, 201–202
   Federal Historic Sites Act, 153
   founding and expansion of, 30
   standards for protecting cultural resources, 205
National Park System, 29
   Antiquities Act of 1906

Department of Defense and Department of Interior, 104
funding for urban parks, 33
Gateway National Recreation Area, 102–106, 109
Independence National Historical Park, 149–173
mission of, 30
strengths of, 33
National Register of Historic Places, 186
Jacob Riis Park, 103–104
Native Americans. *See* cultural groups
New Haven, Connecticut
   research on parks, 43
New York City
   Battery Park City, 199
   Bronx. *See* parks, Pelham Bay Park; Van Cortlandt Park
   Brooklyn, 32, 43–44
      segregation and gentrification in, 60
      See also parks: Prospect Park
   budget crisis, 31
   City Hall Park, 19
   Coney Island, 31
   Lindsay, Mayor John, 31
   Mulberry Bend, 102
   New York Harbor and the Narrows, 29
   Queens, 32
   Rockaways, the, 27, 102, 108, 115, 134, 203
      Breezy Point, 32
      Fort Tilden, 32
   small parks and plazas, 1
   World Trade Center, 2, 202
   See also parks: Central Park; Columbus Park; Union Square; Washington Square; Whyte, William H.
New York City Parks and Recreation Department
   history of, 131–132
   in Jacob Riis Park, 102, 109, 125
   in Pelham Bay Park and Orchard Beach, 32, 134
   politics of, 204
   in Prospect Park, 47–48, 63
   under Henry Stern, 63–64
   under Robert Moses, 27, 108
   See also parks: Central Park
Nixon, Richard M., 30

INDEX

Observational research methods. *See* research methods
Olmsted, Fredrick Law
　Boston work, 24
　designs, 23, 26
　philosophy of, 209–210
　and Vaux, Calvert, 45–46, 200
　See also parks: Prospect Park
Olmsted Brothers, 23
　neighborhood playgrounds, 26
Orchard Beach. *See* parks: Pelham Bay Park

park-associated people. *See* ethnographic resources
Palm Springs, California, 2, 149
Paris, 149
Paulus Hook. *See* Jersey City, New Jersey: neighborhoods
park design
　landscape design in Prospect Park, 45–46, 62–64
　of landscape parks, 20
　romantic influence, 20
　vernacular traditions, 21–22, 24, 35
　　at Independence National Historical Park, 34
　　at Jacob Riis Park, 198
　　in Pelham Bay Park, 131–134
　　in Prospect Park, 37–38, 198, 201
　See also parks: Jacob Riis Park; Jones Beach; Pelham Bay Park, Orchard Beach
park layout, 203
　of Battery Park, 74–77
　of Independence National Historical Park, 153–157
　of Jacob Riis Park, 108–110
　of Liberty State Park, 82–84
　of Orchard Beach, 27
　of Prospect Park, 43–47
park management
　laissez-faire approach, 34, 127
　personal ties to parks, 129–130
　plan for Independence National Historical Park, 151
　at Prospect Park, 39, 47–48, 198
　use of REAPs, 173
parks
　Acadia National Park, 30

Adirondack Forest Preserve, 25
American Beach, 127–128
Arnold Arboretum, 23
Battery Park, 71, 199
　attitudes toward proposed Ellis Is. bridge, 79–82
　description of, 74–79
Big South Fork National Recreation Area, 187
Blackstone Valley National Historic Corridor, 204
Boston Common, 19
Boston Harbor Islands National Park Area, 33, 203
Cape Cod National Seashore, 23, 204
Cape Hatteras National Seashore, 30
Casa Grande, 29
Central Park, 131, 199, 203–204, 210
　Conservancy, 195
　construction expense, 23
　planning of, 20, 22, 203
　restoration of, 195
Chattahoochee River National Recreation Area, 31
Cherokee, Iroquois, and Shawnee Parks, 23
Chicago Forest Preserve, 24
City Hall Park, 19
Colonial Williamsburg, 155
Columbus Park, 102
Cuyahoga Valley National Recreation Area, 31
Delaware Park, 23
Delaware Water Gap National Recreation Area, 204
Ellis Island, 197–198, 203, 209
　ferry to, 69
　management priority, 34
　proposed pedestrian bridge, 69–71
　relationship to New Jersey, 70
　restoration of, 69
　social class and race issues, 95–97, 99, 197
　Statue of Liberty, 197, 203
Fire Island National Seashore, 30, 204
Forest Park, 23
Franklin Park, 23
　user study, 40
Gateway Arch, 30

parks (*continued*)
  Gateway National Recreation Area
    access to, 197–198
    compared to other national parks, 125
    description of, 28–29
    Floyd Bennett Field, 103–104, 108
    Fort Tilden, 32
    historic preservation in, 104–106
    history of, 31–33
    Jacob Riis Park, 101–127, 208
      access to, 197–198
      back-beach areas, 107, 115–124
      bathhouse, 105, 108, 110, 112, 115, 125, 178, 197
      beach bays, 107, 110–115, 125
      cultural groups at, 106, 197
      in decline, 102, 109
      design approach under Moses, 27
      funding and politics, 33
      historical and social background, 108–110, 197
      management in comparison to Orchard Beach, 205
      REAP methodology, 71–74, 106–107
      social setting
        back beach, 116–119
        beach bays, 111–116
    mission, 34, 41
    as a type of park, 28, 203
  Glen Island, 137
  Golden Gate National Recreation Area, 41, 203
  Great Smoky Mountains National Park, 30
  Highland and Seneca Parks, 23
  Independence National Historical Park, 149–173, 186, 203, 206–207
    cultural groups related to, 151–152, 196–197
    cultural representation in, 17
    founding of, 30
    history of, 153
    Liberty Bell, 158, 160–162, 164, 166, 168–169, 172–173, 202, 206
    neighborhoods related to, 151–152
    REAP methodology, 152–157
  Indiana Dunes National Lakeshore, 30
  Jones Beach, 108
    design of, 27

  Lake Itasca State Park, 25
  Liberty State Park, 70
    attitudes toward proposed Ellis Island bridge, 97–99
    description of, 82–86
  Lincoln Park
    cultural groups in, 43
    user study, 40
  Metropolitan Park Commission, 24
    Lynn Woods Reservation, 63
    Middlesex Fells Reservation, 63
  Minuteman National Historical Park, 204
  Mount Royal Park, 23
  Niagara Falls Reservation, 25
  Parque Central, Costa Rica, 6–7
    research methodology in, 41
  Pelham Bay Park
    as example of woodland reservation, 24–25
    history of, 131–138, 204
    management, 24
    Orchard Beach, 127–148, 205, 208
      cultural groups at, 106
      design approach under Moses, 27
      funding, 125
      historical and social background, 131–138
      Hunter and Twin Islands, 132, 136
      Latino visitors, 138–140, 198
      local seniors, 140–143
      methodology, 129–131
      naturalists, 144–146
      vernacular place-making in, 26, 132–134
  Point Reyes National Seashore, 30
  Prospect Park, 37–68, 131, 197, 208–209
    built features, 46–47, 201
    cultural festivals, 55
    design of, 20, 34, 45–46, 203
    drummers, 54–56
    enjoying nature in, 59–60
    history of, 43–44
    management, 39, 47–48
    Peninsula, the, 38
    picnicking in, 57–58
    social class and race issues in, 60–62
    surrounding neighborhoods, 43–44
    uneven maintenance in, 45

INDEX

User Study, 38–40
  methodology, 38
  findings of, 48–53
  user values, 50–53
  with regard to landscape, 62–64
Rittenhouse Square, 20
Salem Maritime National Historical Park, 30
Sandy Hook. *See* Gateway National Recreation Area
Starved Rock State Park, 25
Statue of Liberty National Monument, 30
Union Square, 1
Van Cortlandt Park, 130, 199, 204
  description of, 26
Washington Square, 1
Yellowstone National Park, 29
Yosemite, 25
park types, 203
  commons, 19
  heritage parks, 31
  landscape parks, 20
  national parks, 28–32
    seashores, 30
    recreation areas, 31
  recreation facility parks, 26–28
  state parks and reservations, 24–26
  urban park projects of National Park Service, 30
phenomenological research methods. *See* research methods
Philadelphia, Pennsylvania. *See* parks: Independence National Historical Park; Rittenhouse Square
picnicking. *See* activity-based groups
place attachment, 202
  and identity, 149
  Independence National Historical Park, 149–150
  Jacob Riis Park, 112, 114–115, 120
  Liberty State Park, 89
  Orchard Beach, 127, 138, 139, 141–143, 147
  preservation, 5
playgrounds
  Jacob Riis Park, 105, 109
  Orchard Beach, 134
  Wood Park at Jacob Riis Park, 116–118
  See also park types: recreation facility parks

Prospect Park. *See* parks
Prospect Park Alliance. *See* parks: Prospect Park, management
Public Space Research Group, 3
  work at Ellis Island, 69
  work at Jacob Riis Park, 102–104
  work at Pelham Bay Park, 129
  work at Prospect Park, 38

Rapid Ethnographic Assessment Procedures (REAP), 175, 183–188
  history of, 183–188
  Independence National Historical Park, 150
  Jacob Riis Park, methods, 106–107
  methodology, 188–193
  methods in Battery Park, Liberty State Park, and Jersey City neighborhoods, 71–74
research methods
  cognitive, 176–177
  constituency analysis, 175, 180
  discourse approaches, 176, 179
  ethnoarchaeological techniques, 177
  ethnographic approaches, 175, 178, 205–207
  ethnosemantics, 175, 177, 182–183
  historical, 176, 178
  observational approaches, 175, 177
  phenomenological, 176–178
  qualitative, 175–176
Rochester, New York. *See* parks: Highland and Seneca Parks
Roosevelt, Franklin D., 30
Russians. *See* cultural groups

Saarinen, Eero. *See* parks: Gateway Arch
Sandy Hook. *See* parks: Gateway National Recreation Area
San Francisco, California. *See* Hetch Hetchy Valley; parks: Golden Gate National Recreation Area
Santa Monica Mountains, 31
Sierra Club. *See* Muir, John
social sustainability, 101, 198–199, 207–209
  explanation of, 5
Society Hill, Philadelphia, 196
  See also parks: Independence National Historical Park

South Americans. *See* cultural groups: Central and South Americans
St. Louis, Missouri
   1904 Exposition, 23
   See also parks: Forest Park; Gateway Arch

tourism, 157
traditionally associated people. *See* ethnographic resources
triangulation, 185, 191

underutilized park, 101–126

Van Cortlandt Park. *See* parks
Van Vorst. *See* Jersey City, New Jersey: neighborhoods
vernacular uses, 205
Vietnamese Americans. *See* cultural groups

Western Europeans. *See* cultural groups
wild land parks, 41–43
Wilson, Woodrow, 30
Whyte, William H.
   rules for small urban spaces, 1, 4
   social viability, 101
   urban design principles, 177, 195